THE

OF

SIMON AND SCHUSTER • NEW YORK

MACHINERY
NATURE

PAUL R. EHRLICH

Published by Simon and Schuster
A Division of Simon & Schuster, Inc.
Simon & Schuster Building
Rockefeller Center
1230 Avenue of the Americas
New York, New York 10020
SIMON AND SCHUSTER and colophon
are registered trademarks of Simon & Schuster, Inc.
Designed by Edith Fowler
Manufactured in the United States of America

10 9 8 7 6 5 4 3 2

Library of Congress Cataloging in Publication Data

Ehrlich, Paul R.
  The machinery of nature.

  Bibliography: p.
  Includes index.
  1. Ecology.   I. Title.
QH541.E35   1986   574.5   85-24980
ISBN: 0-671-49288-8

To
Max and Isabell
and
Stanley and Marion
with thanks

CONTENTS

INTRODUCTION

Butterflies, Ecosystems, and People

WE WERE in our ninth or tenth consecutive day of work on a montane butterfly population. The setting was beautiful; flower-strewn sagebrush meadows stretched to a horizon of snow-capped peaks, and a rock wren serenaded us from his habitual perch on an aspen sapling. Only the determined mosquitoes kept the situation from being totally idyllic. Our group was carrying out a large-scale experiment to determine the structure of the population: how much individual butterflies move around and who mates with whom. The results would provide insight into how butterfly populations change in size and evolve. We also thought the behavior of butterflies would prove similar to that of many kinds of plant-eating insects—humanity's most important competitors for food. If so, our results could have direct application in the struggle to feed the human population.

Things were going well; hundreds of individuals had been captured, numbered, and released. Each male butterfly found along a ridge-

11

top had had its sex organs dipped in dry red fluorescent pigment; those on the slopes below had been dipped in green. The dye remained on the organs, and some of it would be transferred to the female during copulation. Captured females, when viewed under ultraviolet light, would thus reveal whether they had mated with hilltoppers or those from foggy bottom.

As four of us were netting, marking, and dipping the butterflies, and recording data, a red pickup truck approached along the ridge road. It pulled to a halt opposite us, and with a friendly wave a local resident called, "Watcha doin'?" Our answer was standard: "Working on range pests for the government." It was an answer that had evolved over the years—and it was more or less true. The butterflies did feed on plants that grew on the range, and our research was partially supported by the National Science Foundation, a government agency. We had long since learned that an answer such as "marking butterflies to see where they move" would be greeted with incredulity; "dyeing the sex organs of male butterflies to see which females they mate with" might induce hysterical laughter.

Finding ways to deal with such casual questions is a problem familiar to field ecologists. Very few people understand why ecologists (and other scientists) often work on systems of no immediate economic importance; fewer still have any acquaintance with the kinds of questions asked by ecologists or the answers they have found. Indeed, while most people are very curious about the functioning of the living world that surrounds and supports them, neither their schooling nor the popular media have provided them with the basic information they need to understand it. And researchers sweating at an altitude of nearly 10,000 feet would not have the time to provide the needed background even if the questioner were inclined to listen to a long explanation. Our visitor, fortunately for us, inquired no further, and we went back to work.

Yet I always feel a little guilty when I duck an educator's responsibility in that way—which is one of the reasons I am writing this book. Nothing is more important to human beings today than understanding how nature works. The very future of our society depends on whether *Homo sapiens* can learn to live without damaging the machinery of nature so seriously that it can no longer support civilization. Therefore, no science, indeed no aspect of human culture, is more important than ecology—the study of the interactions among organisms and between organisms and their physical environments.

There are, of course, very important parts of ecology that normally do not have direct application to environmental issues, even though they are of great intellectual and theoretical interest. After all, ecology is an extremely wide-ranging science. It tries to make sense of the confusing diversity of form, function, and behavior displayed by all the creatures that share Earth with us—an attempt at which it is increasingly successful. For example, ecologists have explained why males of some kinds of birds are monogamous and others support harems. This was a finding of great intrinsic interest but little applicability to the everyday lives of people. But ecology also illuminates the relationship of human beings to the rest of the natural world and investigates the constraints that nature places on human activities. Among other things, ecology provides guidelines for establishing sustainable agricultural systems to support Earth's burgeoning human population—guidelines that are so far being ignored to everyone's great peril.

The basic principles of ecology are accessible to any intelligent person who is willing to put a little effort into learning them. And there are many rewards for that effort. Familiarity with basic ecology will permanently change your world view. You will never again regard plants, microorganisms, and animals (including people) as isolated entities. Instead you will see them—more accurately—as parts of a vast complex of natural *machinery*—as, in the dictionary definition, "related elements in a system that operates in a definable manner." In the chapters that follow, we will look at various aspects of that machinery, from the reproduction and behavior of its individual parts to the principles that govern the functioning of the entire system.

To understand nature's machinery, however, you will need to grasp not only how it operates now, but how it was constructed over billions of years. The process of construction, called biological evolution, has been an off-again, on-again task. The course of construction has been altered by everything from the relative success and failure of individual parts of the ecological machine to catastrophic destruction of entire sections of the apparatus. Evolution has been neither purposeful nor smooth; but it produced us, all of our living companions, and important aspects of our physical environment. In short, we all live in a world that has evolved, and humanity has evolved with it.

Anyone who is not at least passingly acquainted with the history and mechanisms of evolution is out of touch with fundamental aspects of his or her humanity. It is a common separation, and one that an

increasingly hard-pressed *Homo sapiens* can ill afford. And anyone un-
acquainted with evolution will also find many ideas in ecology difficult
to understand. Ecologists considering a structure or a behavior in an
organism try to understand how it evolved. Evolutionary notions are an
integral part of ecological thought; they explain *why* nature's machinery
functions as it does. Thus in what follows we will also examine the basic
features of the process of evolution.

One of the main focuses of this book will be on what ecologists find
to be challenging, interesting, and fun. I will explore a face of ecology
that is unfamiliar to most people. I will not dwell on red-hot environmental issues such as overpopulation, extinction, desertification, acid
rain, toxic waste disposal, whale and seal hunting, energy policy, and
the ecology of nuclear war, although I will mention current issues when
the context is pertinent. Rather I'm going to lead a frankly personalized
tour of the principles of the scientific discipline that, among other things,
provides a basis for formulating and evaluating policies to deal with
those practical issues.

My definition of ecology is, you will note, a broad one. In addition
to standard topics such as how populations change in size and how
ecosystems are organized, it includes evolutionary biology and certain
aspects of behavioral biology. It is a combination of disciplines that is
sometimes called "population biology" (to distinguish it from molecular, cellular, and organismal levels of biological science). But *ecology* is
the best-known term for the combination of disciplines that comprises
the science behind life on Earth; it is also the shortest.

Our tour proceeds from the simple to the complex. It starts with
the relationships of individual organisms to their physical environments
—for instance, how plants harness energy from the sun, how tiny insects
can survive at temperatures that would quickly freeze an unprotected
person to death, and how both plants and animals extract nutrients
from their surroundings. It moves on to the properties of populations—
groups of individuals of the same species (kind) living in the same area.
We'll see how and why populations grow, shrink, and evolve. Then we'll
look at the interactions, such as mating behavior and the formation of
schools and herds, of individuals within populations. That is followed
by an examination of relationships between individuals of different species—lions attacking zebras, bugs eating crop plants, mosquitoes carrying malaria to people, or two species of beetles competing for the same
store of grain.

The properties of communities, which are the assemblages of different species found together in the same area, come next—first, community history and geography (who lived when and who lives where), and then community structure (how and why particular species coexist). We look at such questions as why there are no living dinosaurs and what limits the number of species that populates an oceanic island. And, finally, we explore ecosystems,* the combination of communities and their physical surroundings. This is the ultimate level of ecological complexity, the one that is most often discussed on the political side of ecology. Human beings are embedded in and supported by natural ecosystems, and human beings and human systems are rapidly destroying them. Understanding ecosystems is tantamount to understanding our place in the scheme of things—and saving them is saving ourselves.

This tour of ecology should give you a sense of both the discipline itself and how its practitioners operate—and why I and many other ecologists, like most scientists, enjoy our work so much. The study of ecology can be, for scientist and layperson alike, not just a source of enlightenment or a way of helping to preserve civilization, but a wellspring of entertainment. For knowing how the world works permits more than knowledgeable participation in the great decisions of our day. It also reveals drama everywhere—in back yards and small ponds as well as on African savannas or in the waters over the Great Barrier reefs of Australia. And the plays are free for those who learn how to view them.

Ecologists find pleasure in their work in many ways. Ecology is a broad discipline, and its practitioners vary in their backgrounds, training, interests, and esthetic senses. The greatest pleasure for some is to be in the field, trying to discover nature's secrets. For others, it lies in creating an intellectually satisfying mathematical model that can assist in thinking about a complex natural phenomenon. Still others find great satisfaction in designing meticulous experiments in the laboratory to test raw theory or ideas developed during field observations. Some ecologists are driven by an intense desire to understand how the world works; others by an equivalent desire to apply ecological knowledge for the good of humanity.

This diversity is healthy for ecology as a whole. Because of the immense complexity of many ecological phenomena, there is no guar-

---

* In dealing with a technical subject, some use of technical terms is essential for both brevity and accuracy. I have, however, attempted to minimize jargon and to define each term, either explicitly or by context, when it is introduced.

antee that any one approach will solve a given problem. Many of the most difficult ecological puzzles are likely to be solved only by teams of scientists applying a variety of skills. Therefore the diversity of interests improves the chances that the required skills will be available when needed and encourages a rapid evolution of the discipline.

There are, however, dangers in diversity. The obvious one is fractionation into subdisciplines, or even "schools" within subdisciplines, and a subsequent lack of communication, which can only be damaging to the science as a whole. Ecology has suffered historically from these tendencies and, to a much lesser degree, continues to do so today. Ecologists who study populations do not talk as much as they should to those who are interested in the complexities of ecosystems. Theoreticians sometimes are cut off from fieldworkers and vice versa. Disagreements over difficult-to-resolve questions sometimes become intense, as you will see. Fortunately, however, most disputes have tended to sharpen the science and clarify, if not always settle, issues.

But the major trends in ecology today inspire hope. There is increasing interest in long-term studies, which are desperately needed to answer questions as different as: What causes changes in population size? How important is competition between species in shaping communities? And what is the impact of disturbances on ecosystems? Theoreticians, experimentalists, and natural historians are now starting to listen to one another and work together, to the benefit of all three groups. And, above all, a truly impressive array of bright young minds has been attracted to the field in the last two decades—including, at last, numerous women.

Ecologists have gone a long way toward explaining different aspects of the extremely complex systems they investigate. Now they are beginning to tie the pieces together. Integration cannot come too soon, because a better understanding of ecosystems is essential to saving them. A common question from politicians, for instance, takes the form: "What will be the consequences of the extermination of the snail darter, or of chopping down fifty percent of the remaining Amazonian rain forest?" At the moment, the answer must be given in terms of general concerns rather than relatively assured and specific predictions. Sadly, uncertainty about consequences is all too often interpreted by decision makers as carte blanche for not proceeding until "certainty" can be achieved. This has been amply demonstrated by the lack of political will to abate acid rain until enough research has been done to be "sure" that it is causing serious harm.

Obviously, ecologists and other scientists must try to educate non-scientists about how to evaluate the risks associated with various courses of action, since, technically, there is *never* complete certainty in science (or, indeed, in any area of human endeavor). Consider what may be the greatest ecological risk today—that of a "nuclear winter" following a large-scale nuclear war. As a first approximation, risk can be viewed as the product of the chance of a harmful event occurring multiplied by some measure of the amount of harm it will cause if it does occur. (For example, a 50 percent chance of losing $10 is thus considered about the same risk as a 5 percent chance of losing $100.) In such a formulation, it matters little whether the probability of a nuclear war causing a nuclear winter is 5 or 95 percent. Nuclear winter would so severely damage the ecosystems of the Northern Hemisphere, or perhaps of the entire world, that it would wipe out billions of people and perhaps even exterminate all of humanity. Thus the second factor in the risk equation, the harm, is unbelievably large. That in turn makes the size of the first factor—the probability of a nuclear winter—much less important. The product of the two factors, the risk, remains gigantic even if there is only a 1 percent chance of a nuclear winter.

But decision makers in Western society do not grasp this fundamental principle (if they did grasp it, the nuclear arms race might have ended long ago), and there is little sign that they are suddenly about to discover it. Therefore the future influence of ecologists on policy will be at least partly a function of how rapidly developments in basic ecological science can reduce the uncertainty that impedes political action on acid rain, nuclear winter, and many other environmental issues.

The main hope for changing humanity's present course may lie less with politics, however, than in the development of a world view drawn partly from ecological principles—in the so-called deep ecology movement. The term "deep ecology" was coined in 1972 by Arne Naess of the University of Oslo to contrast with the fight against pollution and resource depletion in developed countries, which he called "shallow ecology." The deep ecology movement thinks today's human thought patterns and social organization are inadequate to deal with the population-resource-environment crisis—a view with which I tend to agree. Within the movement disagreement abounds, but most of its adherents favor a much less anthropocentric, more egalitarian world, with greater emphasis on empathy and less on scientific rationality.

I am convinced that such a quasi-religious movement, one concerned with the need to change the values that now govern much of human activity, is essential to the persistence of our civilization. But agreeing that science, even the science of ecology, cannot answer all questions—that there are "other ways of knowing"—does not diminish the absolutely crucial role that good science must play if our over-extended civilization is to save itself. Values must not be based on scientific nonsense. The notion, recently put forth by a professor of marketing, that resources are infinite because "copper can be made from other metals" cannot be made true by even the most passionate belief. Likewise, human population growth cannot continue forever just because some religious or political dogma holds it to be possible or desirable.

Given the present level of human overpopulation, only a combination of changes in basic values *and* advances in science and technology seems to provide much hope for avoiding unprecedented calamities. And only an appreciation of how nature's machinery functions can provide the basis for the necessary changes in attitude and the guidelines for the safe deployment of new technologies. Developing that appreciation is an infinitely simpler task than converting civilization to a whole new set of values. But once most people understand ecology, that conversion of values should be easier.

CHAPTER 1

SURVIVING IN THE PHYSICAL ENVIRONMENT

Physiological Ecology

ON A SUNLIT summer day in a suburban back yard, or in a flower-strewn mountain meadow with butterflies flying and birds singing, a multitude of interactions is taking place between organisms and their physical environment. Energy is being captured from the sunlight by the plants, water is constantly flowing through them, and all the plants and animals are exchanging gases with the atmosphere. But these processes are not obvious—indeed, they were largely unknown a couple of centuries ago.

If the physical environment changes, though, the relationship between the organisms and their surroundings is often thrown into sharp focus. For example, when night falls, birds are hushed and butterflies stop flying. Or, on a longer time scale, if a freak snowstorm hits a mountain meadow late in June, or a homeowner forgets to water the back yard for a few rainless weeks, the effects are dramatic. The flowers in the meadow are destroyed; the lawn turns brown. Each individual or-

19

ganism is constructed so that it can grow and reproduce in a given set of environmental conditions. If it finds itself outside the boundaries of that set of conditions, its chemical life processes (metabolism) grind to a halt. Too much cold or too little water will often take plants or animals beyond the limits of survival.

Physiologists study the functioning of entire organisms and their parts; physiological ecologists investigate the relationships of that functioning to external conditions, especially those of the physical environment. This is the most basic level of ecology, since its concern centers around the individual organism, rather than the population, community, or ecosystem. It also is of great practical importance. Whether a certain strain of wheat or breed of cow is suitable for use on a farm in a given climate are primarily questions of physiological ecology. So are such questions as what will happen to organisms in a stream if power-plant effluents heat it or if acid rains change its chemical composition.

The questions addressed by physiological ecologists often are suggested by observations of organisms under stress—either those living in habitats ordinarily subjected to extreme physical conditions (such as the daytime heat and lack of water in deserts) or those faced with unusual circumstances (such as a drought or unseasonal frost) in their normally moderate environments.

The impact of physical stresses on animals and plants can be severe —as any tropical-fish fancier who has had an aquarium heater fail during a cold snap can testify. Neon tetras and angelfishes will die if the water in their aquarium drops below 60° F. or so. In contrast, trout will not breed if the water in their streams gets that warm. And goldfishes, which can live in 75° water, will not be in trouble even if the water temperature falls to near freezing, as long as the fall is gradual.

Several important principles lie beneath these simple observations. The first is rather obvious—plants and animals have evolved the ability to deal with the physical conditions to which they are normally exposed. Or, as evolutionists like to say, they are adapted to those conditions. Some species, especially those living in changeable environments such as estuaries or terrestrial areas with distinct seasons, can tolerate a rather wide range of conditions. On the other hand, species that live in fairly constant environments, such as the deep seas, tropical rain forests, or (in the case of some parasites) the inside of another organism, may be able to survive only within a tight envelope of physical conditions.

In geographically widespread species, different populations may evolve different tolerances. For example, European populations of Mediterranean fruit flies (Medflies) can withstand cold that would kill flies of the same species from Hawaii. Ignorance of the geographic origin of the invaders handicapped the 1981 campaign to eradicate the Medfly in California, where it had been accidentally introduced. There was extensive spraying with the insecticide Malathion, and the Medfly population did disappear. But it remains unclear whether the Malathion or a cold winter was the primary reason for its disappearance.

The second ecological principle our observations illustrate is that many, if not most, individual plants and animals can change their tolerances to physical factors if exposed to gradually changing conditions. They can, to one degree or another, acclimate to new circumstances. Anyone who has traveled to the mountains has directly experienced acclimation: exercise that causes extreme panting and distress from oxygen shortage on arrival at high altitude will cause little discomfort after a few weeks in the thin air.

The basis of human acclimation to high altitudes is well understood and results from a number of physical changes. Although we are usually not aware of it, our breathing gradually becomes deeper and more rapid so that air is delivered more quickly to the surfaces in the lungs where oxygen is transferred to the blood. At the same time, the blood supply to the lungs increases, and the capacity of the bloodstream to carry oxygen also expands by means of, among other things, an increase in the number of red blood cells. These cells contain hemoglobin, the oxygen-carrying pigment that makes our blood red. All these changes take place at different rates: the breathing changes are completed within a few days, but full augmentation of the red blood cell supply takes more than six weeks, although an early response occurs within two or three days.

If very long periods are spent at high altitudes, further, slower changes occur in tissues to make them more efficient in the oxygen-short environment. Muscles, for instance, develop more capillaries, the fine vessels that deliver the blood. All of these factors combine to ameliorate or eliminate the original distress.

Acclimation, by the way, must be carefully distinguished from evolutionary adaptation, which is a quite different process. Acclimation occurs in an individual and is not passed on to the next generation. If a woman lived at high altitude for years and then carried and delivered a baby at sea level, the baby would not be acclimated to thin air. Indeed, the mother would tend to lose her acclimation. In contrast, evolutionary adaptation, such as the development of resistance to DDT by insects, occurs in populations over many generations—by mechanisms that I'll discuss in the next chapter. And evolutionary adaptation *is* passed from generation to generation.

A third principle of physiological ecology is that there are limits to both evolutionary adaptation and acclimation. No individual fish or fish species can survive in both the Arctic Ocean and an Amazonian freshwater pond. And nothing can live on the sun. This is because numerous constraints are placed upon organisms by their chemical makeup. For example, various proteins are key components of the physical structure of all plants, animals, and microorganisms. Other proteins, serving as biological catalysts, are essential regulators of life processes. Proteins have three-dimensional shapes that are critical to their functioning, and high temperatures cause their shapes to change. Most proteins cannot retain their structure at the temperature of boiling water, let alone at the temperature of the sun. That is one reason why boiling is such an effective means of sterilization, since bacteria (just like people) die if their proteins lose their shapes.

The living world always seems to hold surprises, however. Strange bacteria have recently been found living and reproducing in superheated water coming from "black smokers"—deep-sea hot springs whose effluent remains liquid (rather than steam) only because of the tremendous pressure at depths of many thousands of feet. The proteins (and other crucial large molecules) of these bacteria remain functional at temperatures of some 600° F., three times the boiling point of water at the sea's surface, and the bacteria can reproduce at about 500° F. The secret lies in the pressures; the outer limits for the existence of living systems seem to be set not, as was once thought, by an appropriate temperature range, but by a combination of pressure and temperature conditions that permit water to exist as a liquid. Perhaps the most exciting result of this extraordinary discovery is that it greatly expands the possible environments where life might be found on Earth and in the universe as a whole.

In spite of the existence of such exotic life-forms, however, the physical conditions within which the vast majority of earthly organisms can thrive are quite restricted. Most plants, for instance, require substantial amounts of sunlight. In fact, all significant ecosystems depend upon energy from the sun, captured by green plants in the process of photosynthesis. There are a few bacteria, such as those found around the "black smokers," that derive their energy from chemical reactions not driven by sunlight. They represent, however, only a minuscule fraction of known life-forms; it is photosynthesis that captures the energy for all familiar plants and animals.

Photosynthesis plays such a key role in the ecology of our planet that I must expand briefly on it and its ramifications here before continuing with topics that are based on a knowledge of it. Photosynthesis is an extremely complicated process in detail, but in outline it is fairly simple. Through photosynthesis, plants are able, with the help of sunlight, to combine carbon dioxide and water, both of which have a low energy content, into high-energy carbohydrates such as sugars, starches, and cellulose. Light energy from the sun is absorbed by green pigments called chlorophylls, which give the plants their green color. That energy from light is then transferred in a complex series of steps to the energy of chemical bonds that hold carbohydrate molecules together. Oxygen is given off as a by-product of photosynthesis.

Both plants and animals are then able to use the energy in those chemical bonds to drive their life processes. Animals, of course, acquire the energy by eating the plants, or by eating plant-eaters, or by eating other animals that have eaten plant-eaters, and so on down the line. The way the energy is obtained from the bonds is basically the same in both plants and animals. Again, the details of the process are complicated, but the principle is simple. The process is known as cellular respiration and consists of a controlled oxidation—a sort of slow burning—of carbohydrates. It is roughly the reverse of photosynthesis: oxygen reacts with the carbohydrates and releases carbon dioxide, water, and some of the energy that originated with the sun. That energy is then harnessed by special chemical compounds containing phosphorus and is used to divide cells, grow leaves, reproduce, flex muscles, think, and do all of the other things that we associate with living.

Not surprisingly, then, green plants are categorized as the producers in an ecosystem. Even though technically they don't produce the energy, they do transfer it via photosynthesis from the physical to the biological part of the system and make it available to all the other organisms, the consumers. Ecosystems normally contain various trophic (feeding) levels of consumers—herbivores that eat the plants, carnivores that feed on herbivores, carnivores or parasites that feed on other carnivores, etc. Feeding sequences, such as corn $\rightarrow$ cow $\rightarrow$ person $\rightarrow$

mosquito; or alga → tiny shrimplike crustacean → little fish → big fish → shark, are called food chains. Food chains are normally woven together into food webs, since, for instance, people can eat corn, fishes, and cows, and sharks occasionally dine on people.

Energy, by the way, is not the only important "product" of photosynthesis. The vast majority of the oxygen in the atmosphere is also a product of photosynthesis; indeed, in the course of geologic time, green plants have completely changed the chemistry of Earth's atmosphere. Oxygen gradually accumulated in the atmosphere over more than a billion years because more was produced by photosynthesis than was used in oxygen-consuming processes, including respiration. In fact, atmospheric oxygen did not reach its present concentration until about 600 million years ago. Only as the atmosphere became oxygen-rich did the physical conditions of Earth's land surface become suitable for occupation by organisms. And only as oxygen became available for the production of ozone was a shield of the latter formed in the upper atmosphere, blocking out harmful ultraviolet light. Then organisms that once had to shelter themselves from ultraviolet radiation by living under water were at last free to invade the land. Thus while organisms are dependent on the state of the physical environment for their survival, they also modify that environment in many ways and often make it more suitable for the support of life.

The critical role of sunlight in the life of green plants is one of the most elementary facts of physiological ecology. Plants and (indirectly) animals depend upon that light for the energy to drive their life processes. Although both plants and animals have some capacity to store energy (in the chemical bonds of glucose, for instance), that capacity is usually quite limited. Active organisms—those that are not in some special resting state (such as a winter-dormant maple tree)—require a more or less continuous supply of energy if they are to remain healthy and alive. A substantial interruption in the energy flow spells trouble, as anyone who has gone without food for a few days or has kept a houseplant in the dark for a couple of weeks can testify.

Recently interest in the possibility of a nuclear winter has illuminated the dependence of familiar organisms on a narrow set of physical conditions, especially the flow of energy from the sun. Atmospheric physicists have projected that a large-scale nuclear war is likely to produce catastrophic changes in the physical environment, at least in the Northern Hemisphere, and perhaps over the entire Earth. Soot injected into the atmosphere from the fires ignited by a large number of nuclear explosions, combined with dust lofted by the blasts themselves, could blot out the sun for weeks. In the absence of the sun's warmth, continental surface temperatures could drop below freezing, even in midsummer.

The darkness would be, all by itself, a recipe for biological disaster. Blocking the flow of sunlight to an ecosystem amounts to cutting off its food chains at their bases. The rate of collapse of the food chains would depend in part on the capacity of organisms in the chains, especially the producers, to store energy. Collapse in the absence of sunlight would be especially swift in marine systems where most of the producers are single-celled algae that float in the water (phytoplankton)—tiny green plants with exceedingly small reserves of energy.

So turning off the light for an extended period would cause what might be thought of as "ecological starvation." Lowering the temperature would also, by itself, pose a severe threat to the survival of organisms that are not tolerant of chilling or freezing.

Tropical and subtropical plants and animals not adapted to the cold would be especially at risk. Even those creatures that live in areas regularly subjected to freezing temperatures, however, such as rodents (which hibernate) and aspen trees (which, in a dormant state, easily survive subalpine winters in Colorado), could suffer severely in a nuclear winter. This is because both plants and animals normally need a period of gradual preparation before they can survive the stresses of winter.

Many trees go through a three-stage process of cold hardening. As days grow shorter, but before the first frost of the season, growth stops and physiological changes take place that prepare the plant for the next stage. Frost then triggers the second stage, and extreme cold induces the third. The changes that occur in all three stages have only been partially elucidated by physiological ecologists. One change involves binding water to proteins, a process that discourages the formation of lethal ice crystals within the cells. In another, delicate cellular machinery is disassembled and stored in simple units that are less subject to cold damage.

Not surprisingly, different kinds of trees harden to different degrees, and the amount of hardening against cold a given tree undergoes is related to the temperature regime to which it is normally exposed. The magnolia of the southeastern United States cannot resist temperatures below about 0° F. in the winter; the northern paper birch can endure temperatures colder than minus 100° F.

When trees from different populations of the same species harden is also related to where they live, since different places have different temperature regimes. The red osier dogwood, *Cornus stolonifera,* is a small tree that occurs in a broad band across the northern United States, subarctic Canada, and Alaska. Individuals of this species from different locations were grown together in an experimental garden in Minnesota, and all were able to survive a temperature of 150° F. below zero in midwinter. But trees from a Seattle, Washington, population could survive a temperature drop to minus 10° F. only after October 17, a month later than dogwoods from North Dakota.

Other evidence shows that there are regular "hardiness rhythms" that occur in woody plants *regardless* of environmental clues. For example, during their flush of growth in the spring, many plants are not capable of hardening under any temperature or day-length regime.

The net result of all this is that trees that can easily withstand 150° F. below zero in midwinter may be killed when exposed to temperatures of less than 15° F. above zero during the growing season. The actual temperature tolerance limits of the trees at any time is much narrower than is implied by the temperature extremes that they can survive in the course of an entire year. This is why the sudden arrival of a nuclear winter during the spring, summer, or early autumn would be likely to kill a great many woody plants. For many species, temperatures in a nuclear winter might drop below even the limits of their winter hardiness. For many others the timing and nature of the environmental changes would not permit hardening to occur. Brief partial thaws, which might occur if the postwar soot-dust veil were uneven, would not make things any better, since they would reverse any hardening that might have occurred. Needless to say, the cold also would finish off growing nonwoody plants—including any wheat, rice, tomatoes, or other annual crops—that the postwar darkness itself had not done in.

Like plants, animals caught unexpectedly by freezing weather often cannot survive. They too tend to have narrow temperature tolerances determined by their physiological status, which in turn is determined by the season of the year. For instance, animals, again like plants, usually pass the winter in a special state of preparedness. Most insects that are able to survive winter in cold climates do so in a "hardened" resting stage.

Some insects harden by emptying their guts and survive because the water in their cells supercools without freezing. Others actually introduce certain fats or sugars into their body fluid to lower the temperature at which ice crystals will form. The best natural antifreeze of all, glycerol, is used by a subarctic scavenger beetle, whose body fluid varies from containing zero percent glycerol in the summer to having almost 24 percent in the coldest part of the winter. When fully cold-hardened, these beetles can survive for a short time at temperatures as low as minus 125° F. But, if they are "surprised" by cold weather, with too little time to synthesize glycerol, a drop to 5° F. above zero is lethal.

Not all insects have special cold-hardy stages, even though they must remain within their range of temperature tolerance. The overwintering stages of many aquatic or soil insects avoid freezing simply by living sufficiently deep in water or soil. Honeybees avoid freezing by "balling"—clustering in a mass and continuously vibrating their wing muscles to produce heat.

The greatest tolerances of cold temperatures are found in warm-blooded animals, many of which can survive and remain active in freezing cold so long as they have sufficient energy available to stoke their metabolic fires. While many birds move south to find abundant food and avoid the rigors of winter, others do not migrate. Tiny chickadees, weighing less than half an ounce, manage to survive the winter in climates as frigid as that of Labrador, the Yukon, and Alaska. They do this by allowing their body temperatures to drop from a normal 104° F. in the daytime to about 86° at night. This lower nighttime body temperature saves precious energy. Even with that saving, however, the chickadees metabolize some 70 percent of their reserve fat in a single night and thus would run out of energy early the next day if they did not feed. So they must start foraging at dawn and continue feeding all day, even in the most miserable weather.

In contrast to chickadees, species such as jays, squirrels, beavers, and human beings prepare for long periods when the physical environment is not productive by storing energy in caches of food—piles of acorns or piñon nuts, branches covered with tender bark, wheat in grain elevators, and so on. Other animals, from lizards to bears, store energy in the form of body fat to see them through such bad times.

The fat-storing small mammals normally enter a long period of torpor with lowered body temperatures to pass the food-short winter. Larger animals, such as bears, are inactive but do not truly hibernate, as ground squirrels do. The key difference is in their winter body temperatures. An inactive bear has a temperature decline of only five degrees or so. Even at that, its energy savings are substantial. In contrast, a hibernating ground squirrel's temperature drops to just a few degrees above freezing, and its heartbeat may slow from hundreds of beats per minute to only five. The squirrel arouses itself about every two weeks, its body temperature returns to normal, and it urinates. It remains active for half a day or a day, it may eat something or defecate, and then it returns to hibernation for another couple of weeks. The reason for these periodic awakenings is not known, but they account for 80 to 90 percent of the animal's energy expenditure during hibernation.

Indeed, physiological ecologists have not yet unraveled many of the mysteries of hibernation—including the exact mechanism that triggers it. Apparently, it is not the onset of cool weather in the fall or the change of day length. If ground squirrels are artificially kept warm and on a constant day length all year around, they still show physiological signs of entering hibernation in their first fall and every year thereafter (although, deprived of day-length clues, their physiological "year" is somewhat less than 365 days). Of course, a squirrel kept in a laboratory chamber at 95° F. cannot lower its body temperature, but it will burn up its stored fat during the winter just as if it were in hibernation. Like many other organisms, the squirrel has an internal biological clock that regulates the hibernation cycle—but how such clocks keep time is an enduring mystery of physiological ecology.

This has been just a sampling of what physiological ecologists have discovered about how animals cope with cold. But even this armamentarium would not protect most of them from the unexpected cold stress of the sort anticipated in a nuclear winter, which would be no kinder to most of the fauna than it would be to the flora. If animals were not prepared with stored fat or cached food, or if they were not in the proper stage of development, they could not survive the unaccustomed cold period. If, like the chickadees, an animal must eat for eight or more hours per day in cold weather just to stay alive, the darkness would make that difficult or impossible. A little basic physiological ecology tells us that a nuclear winter would not only directly threaten any human survivors and agriculture but would also be catastrophic for natural ecosystems. And, as we shall see toward the end of the book, those systems must continue to function properly if civilization is to persist.

Of course, physiological ecologists don't spend most of their time trying to predict the impacts of catastrophic changes like those projected to follow a nuclear holocaust. One of their main tasks is simply to find out how the normal physical environment determines where and how an organism can live. In the process, they gather information on what is going to make life impossible for a given organism. This helps to solve many practical problems, such as which plants are likely to do well on freeway center strips and how far north in the United States invading African "killer" bees will be able to survive. (The answer to the latter question, according to ecologist Orley Taylor of the University of Kansas, is about as far north as Santa Cruz, California, central Texas, and southeastern North Carolina—places where the average daily high temperature for January is at least 60° F.)

Physiological ecologists are also intrigued by puzzles whose answers have no obvious or immediate practical value. For example, closely related organisms often live in quite different microclimates, and physiological ecologists want to know both how they do it and why one moved into a site with one set of physical conditions and the other into another. For instance, one tall tree in the pea family, *Mora excelsa,* thrives as a major component of the solid canopy that forms the upper level of Trinidadian rain forests, while a short member of the same family, *Brownea latifolia,* happily grows underneath it.

This is a bit of a mystery because the differences between the microclimatic conditions in the canopy and those beneath it are quite striking. The most obvious difference is the amount of sunlight available for photosynthesis. Usually less than 5 percent of the light that strikes the canopy gets through to the layers of plants below it. Indeed, the general lack of light greatly reduces the number of plants that can grow beneath the canopy, with the result that, deep in a rain forest, walking around is quite easy. The tangly "jungle" of Tarzan movies is actually not characteristic of tropical rain-forest interiors, but rather of their edges along streams and other places where abundant light can penetrate.

The plants of the understory thus must have a photosynthetic apparatus attuned to functioning at lower light levels than those of the canopy. In addition, they must deal with a different light quality. Much of the light that reaches the deep forest interior is greenish, having already passed through leaves higher up and lost much of the red and blue parts of the spectrum, which are the ones most useful for photosynthesis. Thus understory plants have a special photosynthetic process

ALL PHOTOS ARE BY THE AUTHOR, UNLESS OTHERWISE SPECIFIED.

Tropical rain forest in Costa Rica. Note relative lack of understory, and the buttresses that may function to support the tree in the relatively thin, moist soil.

that operates at relatively high efficiency under those special low-light conditions. Because they are adapted to the filtered light of the rain-forest interior, many understory plants make fine house or office pets—since their often-attractive leaves drop only rarely and they neither need nor do well in strong sunlight, but can thrive in the dimmer illumination of living rooms and doctors' offices.

The amount and quality of light is only one of the microclimatic differences between the canopy and the forest interior. Another is that there is virtually no wind within the rain forest. For that reason, wind pollination would be a poor strategy for understory plants; animal pollination is a superior one. Even members of the grass family, elsewhere a group with largely wind-pollinated flowers, have developed bright floral parts to attract insects in the rain forest. But since, in the semi-darkness of the forest interior, even bright colors are hard to see, strong attractive scents have also evolved. Indeed, many rain-forest plants have small greenish-white flowers that attract flies and small bees with odors.

The ne plus ultra of this strategy is represented by certain species of the weird plant family Rafflesiaceae—thought by some to be distant relatives of the dutchman's-pipe. The Rafflesiaceae are parasites of other plants; they solve the light problem by stealing the products of photosynthesis from their hosts. They grow largely as funguslike strands within the tissues of their hosts' roots and stems and only show their true colors as flowering plants when the time for reproduction arrives. Then the results can be truly spectacular. In the rain-forest species *Rafflesia arnoldii,* the flower is some three feet across and smells of rotting meat; one of its common names is the "stinking corpse lily." While it may not be attractive to humans, the odor lures flies, which carry out the essential task of transferring pollen from male to female flowers.

The lack of wind beneath the canopy helps make the forest interior very humid. This presents the plants that live there with another problem. Their leaves are long-lived; presumably with little light available the plants cannot afford to shed and replace leaves very often. But in their moist environment, the leaves could quickly become covered by a growth of fungi, mosses, and lichens that would cut off what little light they do receive. To retard this process, their leaves have smooth surfaces and especially long "drip tips," designed to shed rainwater fast and discourage the growth of possible settlers.

Intricate adaptations to microclimates can also be found in small animals living beneath the tropical rain-forest canopy. Hummingbirds pollinate flowers of *Heliconia*—plants related to the banana with blooms similar to the bird-of-paradise flower. Within *Heliconia* blooms live tiny hummingbird flower mites that share the nectar with the birds and hitchhike from flower to flower on the birds. My associate, David Dobkin, when he was a graduate student at the University of California, did intensive studies of populations of these mites living in a species of *Heliconia* in Trinidad. He discovered that within different flowers the mites were exposed to different microclimates; flowers that were not well shaded did not support mite populations. Indeed, transient exposure of a flower to between thirty seconds and five minutes of direct sunlight (as a "sun fleck" moves over the dark forest floor) appears to be enough to stress the mites seriously.

Even *within* a flower, the microclimate varies from place to place over time, and the mites move about to remain in the most suitable conditions. David's careful experiments have shown that patterns of sun-fleck movement beneath the rain-forest canopy can be of overriding importance in controlling the size of hummingbird flower mite populations. Similar work by him and others on the effects of microclimate on the dynamics of herbivorous insects indicates that understanding the immediate physical environment of small creatures that we consider pests can greatly help in their control.

Microclimates with too much moisture and too little solar energy create problems for plants of the deep forest. In contrast, desert species must cope with too little water. Perhaps the strangest of the plants that have solved this problem is a distant relative of conifers, with the unwieldy name of *Welwitschia mirabilis,* which lives in the fog deserts of southwestern Africa. *Welwitschia* individuals are low-growing, live partially covered by sand, and consist of a thick root-trunk that produces only two long, wrinkled leaves which often split lengthwise. The plant takes about a century to reach full size, and the same leaves (often a yard wide) last for its entire life, wearing off at the ends and growing at the base. Most of the water it needs does not come in rainfall or from the desert sand but is absorbed directly from the fog and dew which is common in the desert areas where it lives. Because it grows so slowly, the species is extremely vulnerable to human disturbance; it is now one of the few protected plants in Africa.

The question of where and how organisms live leads to a central principle of physiological ecology which can be summarized simply as "there's more than one way to skin a cat." Faced with the same problem, similar organisms often evolve different solutions to it. Confronted by the stresses of winter, some birds migrate to warmer climes; others stay in the cold and forage continuously. The former face a wide array of hazards on their migration; the latter often live on the brink of starving or freezing to death. Either strategy, however, can be very successful.

Animals aren't the only examples of cat-skinning strategies, however—plants exhibit the principle as well. While moisture is usually in short supply for desert plants, light, as I've just noted, normally is not. Elegant studies of their physiological ecology have shown, however, that different plant species have adopted quite dissimilar strategies for taking advantage of the abundant sunlight. My Stanford colleagues Hal

Mooney and Joe Berry, along with Jim Ehleringer from the University of Utah, compared the photosynthetic mechanisms of two annual plants that grow in Death Valley after winter rains. One, *Camissonia clavifor-mis* (evening primrose family), has leaves that are fixed in position. The other, *Malvastrum rotundifolium* (cotton family), has leaves that change position throughout the day so that their blades are directly facing the sun.

Camissonia has an exceptionally high photosynthetic capacity. It achieves its photosynthetic efficiency by allowing carbon dioxide to pass easily into its leaves and by producing a high concentration of certain enzymes needed for photosynthesis (normally there is a shortage of those enzymes, which limits photosynthetic efficiency). The optimum temperature for operation of the photosynthetic apparatus of *Camissonia* is about 70° F., which, allowing for the cooling effect of water evaporating from tiny openings in its leaves, is just about the average temperature of the leaves.

Malvastrum, on the other hand, does not invest as much energy in enzyme production and therefore has a lower maximum photosynthetic capacity. It compensates for this by tracking the sun with its leaves, thus being able to reach its lower maximum level of photosynthesis earlier in the day than *Camissonia* and to maintain it longer. Because its leaves are always facing the sun, they reach higher average temperatures—around 85° F. In response, *Malvastrum*'s photosynthetic apparatus has evolved to operate most efficiently at around that temperature. And, in spite of their very different strategies for doing so, both plants capture about the same amount of the sun's energy per day.

Plants thus face and solve a whole series of problems in acquiring the energy (sunlight) that they require. But they also need materials (carbon dioxide, water, and nutrients) in order to carry out their life processes. Physiological ecologists have elucidated the intertwined ways which many plants have evolved to meet their needs for energy and materials. Water plays a key role in the processes, beyond its direct participation in photosynthesis.

The leaves of most plants have very little structural strength and are kept turgid by the pressure of water in their cells. Since plants photosynthesize, however, the leaves must have large numbers of tiny pores through which carbon dioxide can enter—pores that lead to the moist interior of the leaf. Inevitably, as the carbon dioxide enters, water is lost. Thus to maintain the shape and orientation of its leaves while the pores are open, the roots of these plants must continually take water from the ground and pass it up their stems to the leaves—where it is eventually lost through evaporation from the tiny pores (the process of loss is called transpiration). This is often called the "water cost" of photosynthesis.

The amount of water moved from soil to air by plants is enormous. A single corn plant with a dry weight of a pound at maturity has had something on the order of sixty gallons of water pass through it in its lifetime. Anyone who has failed to water house or garden plants knows about the wilting that results from insufficient access to water; if the roots cannot extract enough water from the soil, the leaves of many plants cannot retain their shape.

Other plants that are adapted to drier conditions are often unable to afford such rapid transpiration and have evolved firmer structures that do not require water pressure for rigidity, such as the stems of cacti, on which to present their green photosynthetic surfaces to the sun. In this way they reduce the flow (and loss) of water. Many desert plants further reduce the loss of what water they do have by means of a special kind of metabolism that permits them to acquire carbon dioxide during the night to be used for photosynthesis the following day, thus keeping their pores closed during the heat of the day, when the potential for water loss is at its greatest. And, interestingly, the water-saving metabolism is not always fixed; some plant species have the ability to switch to the water-conserving metabolism (which is generally less productive) only when water is in short supply. Some loss of water is inevitable, however. No plant has yet evolved a mechanism for taking in carbon dioxide from the atmosphere without giving off water (although plants do not lose water when they use the carbon dioxide produced by their own respiration—a source which is significant in some circumstances).

Animals too have evolved various ways of conserving water when necessary—having an impermeable skin and producing very dry feces are two common ones. And, like plants, they have also become very efficient at extracting energy from their environments. For example, they may specialize in eating certain kinds of organisms, or seek their food in certain special ways or places. Caterpillars of checkerspot butterflies in Colorado consume plants in the snapdragon and honeysuckle families. In the same places, caterpillars of the butterflies known as "blues" usually eat legumes (which are not touched by the checkerspots), while caterpillars of "whites" attack members of the cabbage family (which are avoided by both checkerspots and blues). Each kind of butterfly has evolved the ability to locate its special food plants.

Similarly, some North American warblers often catch insects on the wing, others specialize in gleaning them from foliage, and others in digging them out from under bark. Lions sprint at their prey from ambush; Cape hunting dogs pursue prey relentlessly for long distances. Each of these hunters is structurally and behaviorally programmed to forage for food, and thus acquire the energy it needs, in its own way.

Learning can play a key role in how animals forage. For instance, it is likely that one of the reasons that rats are such successful (if uninvited) companions of *Homo sapiens* is their ability to adjust to novel foods and thus tap novel sources of energy. They do this with a combination of curiosity, a highly developed sense of taste, and great caution. A rat will thoroughly sniff an unfamiliar substance, and, if all seems well, eat a very small amount. It then will not touch the new food for about twenty-four hours. If, after that time, it has suffered no ill effects, it will sample a bit more, gradually adding the food to its diet. If it does feel any discomfort, it will forever shun the food—even if the discomfort was not caused by the food (experiments have shown that rats injected with a substance that causes the same symptoms as food poisoning will thereafter avoid a novel food consumed hours before). This sort of "one-trial" learning makes rats extremely hard to poison with anything that gives prompt symptoms. That is why warfarin, which works by slowly destroying the ability of the blood to clot, is the most successful rat poison.

The dietary habits of people and rats have a lot in common. Rats, like people, prefer to feed in groups—presumably the presence of other rats stuffing themselves reinforces the idea that the food is safe. Both people and rats eat a wide variety of foods, and learning plays a major role in what any individual selects. Rats apparently can learn about a novel food by detecting the odor of the food on other rats. Dietary preferences, once established, change only slowly—as is evident in people from the way immigrants always carry their native cuisine with them to a new home. People, like rats, also often become averse to foods that they have eaten prior to a stomach upset. Both rats and young children, when presented with an array of natural foods, will select from them a balanced diet. In both species, part of the basis of early choice may be a preference for the mother's diet, transmitted by flavors in the mother's milk.*

Energy, oxygen, carbon, and water are vital resources that virtually every plant and animal must acquire. But they also need some twenty other elemental nutrients, including large amounts of nitrogen, potassium, and phosphorus, as well as moderate amounts of calcium, magnesium, sulphur, and iron. In addition, plants and animals require a variety of trace elements—ones needed only in very small amounts—such as copper, zinc, and boron. Plants and animals have similar nutrient requirements because the metabolic machinery of all higher organisms is of the same general design and made up of similar chemical constituents. Sulphur, for example, is an important element in all proteins (which otherwise are made up mostly of carbon, hydrogen, oxygen, and nitrogen), and iron is a key constituent of hemoglobin, which carries oxygen in animals, and of molecules involved in the energy-handling apparatus of both plants and animals.

Except for oxygen, animals normally obtain the nutrients they require from their food and water. But since plants do not forage and, with a few carnivorous exceptions, do not eat prey that contain the substances they need, they must extract nutrients from their physical environment. These are usually obtained from the soil by the roots of the plant (or directly from the water, in the case of aquatic plants).

Of course, if plants continually "mined" soil and water for nutrients, these substances would soon be virtually exhausted (since they are released from rocks only very slowly by physical breakdown or "weathering"), and life would nearly grind to a halt. Fortunately, a mechanism for restoring nutrients to soil and water exists. It consists of a little recognized but essential group of organisms known as decom-

---

* Our primate ancestors ate mostly fruits and other vegetable foods, supplemented with meat whenever it was available. As biologist Tim Roper, an expert on the diets of rats and people, put it: "We are left with the frugivore's passion for sweetness, and the carnivore's passion for animal meat and fat." These evolved preferences were adaptive for most of our history, when starvation was a much more likely prospect than obesity. But now, as Roper points out, Western society is supplied with an abundance of sweets, meats, and fats, all of which are not only attractive for good evolutionary reasons, but also because they are heavily marketed. Given unlimited access to "junk foods," both human beings and rats tend to overeat and become obese. While both species can forage satisfactorily on relatively unprocessed foods, their evolutionary background betrays them in a nutritional environment that has been dramatically and rapidly altered.

posers. These are mostly obscure creatures such as bacteria, fungi, soil insects, and worms. Decomposers make their living by digesting the wastes and dead bodies of other organisms. They break down laboriously created organic molecules into their simple chemical constituents and return them to soil (or water) from which plants can reacquire them. My description of food chains early in this chapter was thus incomplete. A decomposer trophic level is connected to every link in all of them— the decomposers complete the nutrient loops that allow nature's machinery to keep running.

Neither the process of nutrient extraction from soils by plants, nor the decomposition process, nor the structure of soils is simple. Fertile soil is in itself an astonishingly complex, stratified ecological system, containing not just small fragments of the parent rock, but a special dark substance called humus (a mixture of decaying organic matter and the microorganisms that are decomposing it) and a rich flora and fauna. For example, beneath a square yard of Danish pasture, 10 million roundworms, 45,000 small relatives of earthworms, and 48,000 minute insects and mites were counted. A census of 1/30 of an ounce of soil from a fertile farm has turned up 30,000 protozoa, 50,000 algae, 400,000 fungi, and more than 2.5 billion bacteria. And the chemistry of the soil system is extremely complicated, governed by such things as the electrical charges that occur naturally on clay and humus particles.

To underscore the importance of soil organisms, I need only point out that the truly dominant organisms in many forests are not the trees but mycorrhizal fungi that live in the soil. Mycorrhizae are mutually beneficial associations of roots and particular fungi, as experiments with white-pine seedlings have shown. White-pine seeds can be sprouted and the seedlings grown in a sterile nutrient solution for a couple of months. If they are then transplanted to certain prairie soils, they will die of malnutrition. But seedlings grown first in forest soil and then transplanted thrive. The reason is that, unlike their sterile-grown relatives, the seedlings grown in forest soil carry with them the mycorrhizal fungi, which penetrate or surround the cells of the plants' roots and aid them in the uptake of nutrients (in return, the fungi receive energy in the form of carbohydrates).

In addition to the mycorrhizal fungi, the collection of organisms in the soil system around roots is greatly modified by the presence of the plant, which adds organic matter to the soil. Around the roots of spring wheat, the concentration of many kinds of microorganisms is two to twenty times as dense as in soil away from the roots, and at least two kinds of bacteria are more than a thousand times as dense. Many of these organisms (in addition to the mycorrhizal fungi) are directly involved in helping the plant itself take up minerals.

Soil thus does far more than prop up plants; it is a delicate system critical to plant growth and thus to the health of entire ecosystems. A wide variety of insults, including erosion, misuse of fertilizers and pesticides, acid rains, and fires, may alter the chemistry of the soil ecosystem, decimating its biota (all of its organisms collectively) and compromising its plant-nurturing capacity. Agricultural economist Lester Brown was not exaggerating when he wrote: "Civilization can survive the exhaustion of oil reserves, but not the continuing wholesale loss of topsoil."

We have just seen examples of another important principle in physiological ecology: that organisms can shape the physical environment of other organisms. Tall plants change the quantity and quality of light that bathes short plants, decomposers add mineral nutrients to the water taken up by the roots of all plants, and so on. Now let's look closely at the role of the physical environment in shaping organisms—specifically, how evolution tends to produce similar organisms where conditions are similar.

A striking example is the milkweeds growing in African deserts and the cacti of the Americas. The milkweeds look like cacti, but belong to a totally different group of plants. The similarities of the two groups of plants are the result of what is known as convergent evolution—the independent appearance of similar features in unrelated organisms living in similar habitats. The convergence here includes both superficial characteristics and a fundamental physiological attribute, since both groups of plants share the special water-conserving metabolism described earlier.

Other examples of convergence are found in the mammals in Africa and South America. The capybara, a South American semiaquatic social rodent, may weigh more than 150 pounds. It is very similar in shape to the African pigmy hippopotamus, although the latter is two to three times as heavy and belongs to a totally different evolutionary group. Another large South American rodent, the paca, which may weigh twelve to twenty pounds, is strikingly similar in shape and weight to the African water chevrotain, a close relative of deer. Both the paca and chevrotain are largely nocturnal and live in deep forest near water. They share a light-spotted camouflage pattern that makes them difficult to see in a sun-flecked forest environment.

The remarkable similarity between the ant- and termite-eating pangolins of Africa and Southeast Asia and the giant armadillos of South America also turns out to be the result of convergent evolution. Because the pangolin is toothless and armored like the armadillos, it was long classified with them. But detailed studies have shown that the pangolin is a member of an entirely separate group. This is thus a classic example of evolutionary convergence due to a common life-style of burrowing and digging after insects.

Of course, the most commonly cited example of evolutionary convergence in response to the physical conditions of a shared environment is the similar shape of whales and fishes (and, in the age of dinosaurs almost 200 million years ago, of some marine reptiles called ichthyosaurs, which looked stunningly like modern porpoises). The reason for this particular convergence is that only a limited number of shapes allows efficient movement through water. Thus the familiar "fish shape" has appeared independently several times in the course of evolution. Here, as with the other cases of convergence mentioned, the morphology (form) and physiology of organisms respond evolutionarily to the characteristics of their physical environment.

Convergent evolution can extend to entire communities as well. Where temperatures, humidities, and soils are similar, biotas tend to resemble one another—similar appearance and/or behavior having evolved in taxonomically diverse groups exposed to the same physical habitats. Hal Mooney has spent a substantial portion of his career doing careful investigations of the similarities and differences of the vegetation among ecosystems with a Mediterranean climate—regions with mild west-coast marine climates, characterized by winter rains and summer drought. Martin Cody of the University of California at Los Angeles has done fine work on the ecology of bird communities in the same ecosystems—especially in comparing the California chaparral system with the matorral of Chile.

The vegetation in these areas, and in the Italian macchia, French maquis, and South African fynbos, is strikingly similar in superficial appearance, consisting of dense stands of evergreen trees and shrubs with firm, usually relatively small leaves. (The thick vegetation in southern France provided cover for resistance fighters during the Second World War, and thus gave them their nickname of "Maquis.") All of these Mediterranean-type plant communities are subject to frequent fires. In fact, the fires appear to be important in maintaining the characteristics of the system, since the plants in them have evolved in an environment subject to periodic burnings. Plants adapted to such environments often have the ability to sprout from stumps, and their seeds will not germinate until exposed to intense heat or charcoal. People who build homes in areas of chaparral should remember this—it would not *be* chaparral unless it burned over every so often.

Mediterranean ecosystems also show great similarities in their bird communities; each have species of much the same sizes, shapes, and behavior. Like the cacti and desert milkweeds, some of the birds represent true cases of evolutionary convergence, although convergence is often difficult to detect in birds, because the constraints of flight permit only a limited array of shapes in any case.

Physiological ecology has gone a long way toward uncovering the mechanisms by which individual organisms adapt to the physical conditions of their environments and by which they acquire resources. Some of its practitioners who are focusing on the different strategies that have evolved for solving similar problems are shifting their attention from *how* organisms do things to *why* they do them in a particular manner—from mechanistic to strategic reasoning. Why do some plant species flower, produce seeds, and die in a single growing season, while others in the same environment flower and succumb after two years of growth? Why do most species of damselfishes on the Great Barrier reefs abandon their eggs and young to the vagaries of oceanic currents, while mated pairs of one species carefully herd and protect their offspring? Are there general rules for determining how an individual plant or animal should allocate its resources between growth, maintenance, and reproduction? When is it better to adopt the strategy of the elephant— grow big, live long, reproduce slowly—and when that of the mouse— grow small, be short-lived, reproduce fast?

Strategic reasoning could lead to a general solution of one of the most vexing problems in physiological ecology: why some organisms have been able to evolve solutions to certain problems when their close relatives have not. For example, many butterfly species have the ability to ride out unsatisfactory conditions (winter, dry season) in a genetically programmed resting condition called diapause. Different species go into diapause as adults, as eggs, as caterpillars, or as chrysalids. But some closely related species fail to diapause at any life stage. A few, for example, invade temperate North America from the south in large numbers each summer, but are killed back by cold weather each fall.

Such massive mortality among the invaders should constitute powerful pressure to evolve a diapause mechanism. Why diapause has not evolved in those species is a mystery that challenges ecologists, as will be evident as you become more familiar with how populations change their genetic constitution through time. One can answer "how" questions in ecology without reference to evolutionary theory, but for answers to the "why" questions, we must turn to the domains of Thomas Malthus and Charles Darwin.

THE DOMAINS
OF MALTHUS AND DARWIN
Population Ecology
and Evolution

ON THE EDGE of Stanford University's giant campus, in the foothills of California's Outer Coast Range, is a two-mile-long linear accelerator. This valuable tool for investigations in high-energy physics has helped people to understand the structure of the universe. Driving west toward the low mountains on Sand Hill Road, one can see the accelerator on the left. It underlines a fine view of another of Stanford's important research facilities, the Jasper Ridge Biological Preserve.

Stanford is the only world-class university to set aside a large tract (about 1,200 acres) of relatively undisturbed terrain for biological research right on its campus. Generations of students and faculty have worked there, and, if the inventions of the physicists don't bring civilization to an end, future ecologists should be working there long after an earthquake has converted the two-mile-long linear accelerator into four half-mile-long accelerators.

My first visit to Jasper Ridge was in the fall of 1959, shortly after joining the Stanford faculty as a beginning assistant professor. I wanted to start some long-term research there on the ecology and evolution of butterfly populations, because I thought that butterflies provided an ideal system for understanding nature's machinery. I also chose butterflies because I had developed an early interest in these beautiful creatures, and serendipitously they turned out to have characteristics that made them excellent tools for scientific investigation.

All scientists make such choices because they must sample nature; they cannot research everything at once. Scientists study special cases in order to make general points. A system is chosen for investigation because it has features that make it amenable to answering the sorts of general questions posed. For example, in their gradual unraveling of the mysteries of heredity, geneticists have concentrated much of their efforts on fruit flies *(Drosophila)* and gut bacteria *(Escherichia coli)*. These are to most people considerably less esthetically pleasing creatures than, say, butterflies, birds, tigers, or whales.

But the ease with which the flies and the bacteria can be manipulated in the laboratory and the rapidity with which they reproduce make them ideal for working out how "like produces like." And the genetic mechanisms by which tigers produce tiger cubs and killer whales produce killer whale calves turn out to be very similar to those by which fruit flies reproduce themselves. In fact, the mechanisms by which people produce human babies are similar in very important ways to those of other animals and even of bacteria (although it is only the most basic processes that are the same in the latter case). The special sheds light on the general.

Butterflies, in comparison to, say, antelopes or birds, have many of the advantages of fruit flies—they are relatively easy to raise and manipulate in large numbers in the laboratory. And, unlike most insects (but like most antelopes and birds), they are easy to identify and observe in the field. They are also quite short-lived animals and presumably have relatively little capacity for thought or feeling. If individuals must be constrained or killed in the course of research, butterflies pose less of a moral problem for me than would, say, birds or mammals. Above all, the beauty of butterflies, like the beauty of birds, has attracted the attention of large numbers of amateur naturalists. As a result, knowledgeable amateurs have contributed a great deal to what is known about butterfly taxonomy (how to tell the different kinds apart), distributions, and habits.

An adult male Bay checkerspot butterfly taking nectar from sheep parsnip on Jasper Ridge.

In the fall of 1959, a local butterfly collector told me that a population of the Bay checkerspot, a subspecies of Edith's checkerspot butterfly *(Euphydryas editha),* lived on Jasper Ridge. He went with me to the preserve and showed me where a new generation of checkerspots would fly each spring; from parts of its habitat one could see San Francisco Bay, for which the subspecies *(E. editha bayensis)* was named. The flight period is in March and April, toward the end of California's rainy season, so I made plans to carry out a mark-release-recapture experiment then to determine the size of the butterfly's 1960 population and to find out how far individuals moved. Little did I realize that I would be embarking on a research project in population ecology and evolution that would span more than a quarter of a century and involve hundreds of people.

The study of populations of nonhuman organisms was a divided discipline in those days, badly in need of reunion. Some ecologists were interested in the dynamics (changes in size) of populations. Others were interested in how the genetic characteristics of populations change through time—that is, how they evolve. Unfortunately these two groups largely carried on their research separately, with little communication between them.

How and why population sizes change is not just of theoretical importance; understanding the dynamics of populations is of enormous practical significance. Human beings are very concerned with reducing the population sizes of organisms that they consider enemies, from pests that attack crops to parasites that cause illness. They are also interested in enhancing the population sizes of organisms they consider desirable —game animals, crop plants, harvestable fishes, endangered species, and so on.

Understanding genetic change is also more than an intellectual exercise. The evolution of resistance to pesticides in crop pests and disease-carrying animals and of resistance to antibiotics in dangerous microorganisms are two major problems faced by *Homo sapiens.* (At this very moment, both kinds of resistance are involved in the resurgence of humanity's most serious disease—malaria.) And, of course, how the populations of *Homo sapiens* change in both size and genetic characteristics is also of no small moment.

But, even a quarter century ago, it was clear that the dynamics and genetics of populations are not dissociated from each other. Changes in the size of a population influence the evolutionary path that it takes, and genetic changes within a population influence its size. For example, the smaller a population gets, the more likely it is that random events can change its genetic makeup. That, in turn, can increase the possibility that it will become extinct by making the population less able to adapt to environmental changes. On the other hand, the evolution of resistance to penicillin in a pathogenic bacterium can initiate a population explosion of that bacterium and cause the death of its human host.

The unfortunate separation of studies of population dynamics and population genetics was one factor that led me to launch the long-term studies of both the ecology and evolution of checkerspot butterflies. I wanted to see if the two types of studies could be melded to give a coherent picture of how populations behave in nature. And I knew that, difficult as the task might be, I would have fun trying. Checkerspot butterflies would, I hoped, become the fruit flies of population biology —a special case that would illuminate general principles governing the way nature functions and in turn help humanity to solve the many practical problems it faces in dealing with populations of other species. We will thus explore the principles and questions of population ecology by looking first at the dynamics of populations, using the checkerspot butterflies as our primary example. Later in the chapter we will shift our

Aerial view of Jasper Ridge showing the island of serpentine grassland within the chaparral. Lightest areas are yellow wildflowers growing in area C (upper right) and area H (lower left). Area G is too
small to pick out.

focus from how and why populations change in size to what all ecologists know to be a closely related question: how populations evolve.

Early in the spring of 1960, I started going out to Jasper Ridge
armed with a butterfly net, data sheets, indelible felt-tip markers, and
an excellent assistant, a young graduate student named Susan David-
son. The butterflies, I had been told, were found in an area of serpentine
soil. That soil, a type with an unusual mineral content and a characteristic flora, underlay a large portion of the grassland. To help me keep
track of the movements of the butterflies, I arbitrarily divided the serpentine areas of the grassland on the top of the ridge into eight areas,
A through H.

On our first two trips, we did not see any of the checkerspot butterflies. But at 10 A.M. on March 31, on a day that my ancient data sheet
tells me was "sunny—rel. still," I captured my first Bay checkerspot, a
male in area G. It was a lovely creature, faithful to the name "checker-

spot," with a variegated pattern of black, yellow, and red spots, and possessing a sheen characteristic of butterflies freshly emerged from the chrysalis. But its beauty had to be sullied in the name of science. It was held temporarily in an envelope. Then a smear with a felt-tip marker was placed on the underside of the left front wing at the tip to christen it "number 1," and it was released. Next a female was taken in area E, and a smear was placed in a similar position but on the outer edge away from the tip—the place designated to represent the number 2. Our study was under way.

Whenever my teaching schedule permitted, Susan and I were on the ridge, netting checkerspots and gently putting them one by one into envelopes. When we were finished catching in each area, we took each of the captured butterflies out, marked a coded number on its wings, recorded the area of capture, sex, and condition for that numbered individual, and then released it where it was caught. All the original data sheets are still in a file in my office (scientists rarely throw out data), and in reviewing them I see that in eighteen days we handled 206 individual checkerspots.

Of course, we soon found that some of the butterflies that we captured were actually recaptures—they already had our coded numbers on their wings. All of this was dutifully recorded, and the data provided us with the basis for determining two very important pieces of information. The first was the approximate size of the population—this is the classic reason for doing a mark-release-recapture experiment. The second would give us an understanding of the structure of the population —that is, the pattern of movements of individual butterflies.

In principle, determining the first piece of information, population size, from mark-release-recapture data is very simple, as an example will show. Suppose on one day you mark and release 10 butterflies. On the next day you catch 20 butterflies and discover that 2 of them (10 percent) carry marks. This gives an estimate that 10 percent of the butterflies present on the first day were marked. Since 10 were marked on the first day, about 100 butterflies must have been out there. If, on the other hand, only 1 marked individual was found in the second day's sample of 20 (5 percent), the first day's population size would be estimated to be about 200 butterflies. Finally, if 5 marked individuals were found in the sample of 20 on the second day, the first day's population estimate would be about 40 (since 10 is 25 percent of 40).

The basic idea is to discover how much a known number of released marked butterflies is diluted in the population as a whole. But unfortunately, things are not quite as simple as that. All we can get is a *rough* estimate of the size of the population using mark-release-recapture techniques. This is because a lot of assumptions went into the population-size estimates I made in those simple calculations. For example, two important assumptions are that new butterflies do not emerge from their pupae and join the population between the first and second days, and that no butterflies die between the two samplings. We *know* that these assumptions are usually incorrect, and ways must be found to compensate for this recruitment and loss. Although the basic idea is simple, the analysis of mark-release-recapture data can be extremely complex in practice. As a result, the estimates are often only "ball park" ones, and we actually pay attention only to rather large changes in estimated population size since the procedure is not refined enough to accurately detect small changes.

The second important kind of information that came out of our mark-release-recapture studies was how little the butterflies moved around. We quickly discovered there were three separate populations on Jasper Ridge. One was located in areas A through D (subsequently lumped together and called C), a second in area G, and a third in area H. There was very little movement of individuals between these three populations—butterflies marked in areas C or G, for instance, were only rarely recaptured in area H. The butterflies did not venture from one area to another even though the serpentine grassland "island" in which they lived was virtually continuous. One could walk from area to area without meeting even a barrier of low bushes. The butterflies simply did not exercise their obvious powers of travel; they were restricted by "intrinsic barriers to dispersal"—choosing to remain where they emerged from their pupae. Small birds often show similar behavior, not crossing small areas of unsuitable habitat such as a river cutting through a rain forest, even though they clearly are capable of doing so.

Partly as a result of our studies of checkerspots, we now know that understanding the structure of populations is crucial if you wish to influence their dynamics. For example, the best control strategy for an insect pest which is found in many small, isolated populations will be very different from that for one that consists of one or a few giant, widespread populations. The same holds for the best strategy for harvesting a sustainable yield from a fishery or for preserving species of endangered whales.

The butterflies migrated between the populations so rarely that their movements could not have significantly changed any population's size. Therefore we can refer to the three populations as "demographic units," to indicate that they have entirely independent dynamics, although the amount of movement that does occur between units may unite them genetically. Knowing that there were three such units on Jasper Ridge allowed us to calculate separate population-size estimates each year for areas C, G, and H. To my surprise, the population sizes of each demographic unit showed quite different patterns of change. During the first few years, the unit in area H underwent a population explosion increasing its size twenty-five-fold. In contrast, the unit in area C fluctuated in size, and that in area G, located between C and H, declined to extinction.

We knew *what* had happened, but not *why* it had happened. Discovering the reasons for these different dynamic patterns proved a considerable challenge.

I knew that these changes in size in the Bay checkerspot populations, just like those in populations of any other organisms, could be explained in terms of inputs and outputs. The inputs are measured by the rates of birth (or, more technically, natality, since one does not ordinarily think of bacteria, plants, or caterpillars as being "born") and immigration; and the outputs by the rates of death (mortality) and emigration. In order to understand the details of population dynamics, ecologists often construct complicated mathematical models—but the underlying processes are simply these four inputs and outputs.

Since there is very little movement between areas C, G, and H on Jasper Ridge, we can conveniently ignore immigration and emigration in studying their dynamics—just as they can be ignored in analysis of the dynamics of the global human population, because people are not entering or leaving Earth. So one need only consider natality and mortality in studying the dynamics of these populations. If natality is greater than mortality, the population grows; if mortality is greater, it shrinks; and if they are equal, the population stays the same size.

That there were different trends in the three Jasper Ridge demographic units in itself gave us some information on the factors responsible for the changes in population size. It indicated immediately, for example, that changes in the climate of central California were not directly causing changes in either natality or mortality—since all three demographic units obviously were exposed to the same overall weather conditions. We knew, therefore, that we would have to look elsewhere for causes of the observed changes—perhaps, for instance, to ways in which the same weather patterns might have different impacts on the butterflies in areas C, G, and H. But before we consider the answers to this question, let me give you a little more background on population dynamics—both of the Jasper Ridge checkerspots and of animals in general.

First note that the independence of the size changes in C, G, and H confirmed that the demographic units were indeed isolated from one another, since G went extinct just as the adjacent H population underwent an explosion. That first brought to our attention the great importance of understanding the structure of populations if one is interested in determining what controls population size. The identification of three demographic units on Jasper Ridge was crucial. If the grassland island had been treated as a single unit in our mark-release-recapture analysis, rather than having been subdivided, the first few years of the 1960s would have shown a relatively constant population size on the ridge as a whole, and the extinction event would have gone unnoticed. Our understanding of the dynamics would have been flawed, and in an important way. This underlined for us the need to properly identify demographic units.

It is now apparent that the need to identify demographic units is not confined to esoteric studies of butterflies. It can be seen, for example, in the history of the Peruvian anchoveta fishery, which around 1970 supplied as much as 17 percent of the world's harvest of marine fishes. But the anchoveta yields declined dramatically in the early 1970s and have never really recovered. Part of the reason for the decline (or failure to recover) appears to be that harvesting strategies were designed in the absence of knowledge of the number and size of anchoveta demographic units. Similarly, ignorance of the structure of whale populations hinders the development of systems of sustainable whale exploitation (although, on ethical grounds, I favor no exploitation of those extraordinary, intelligent creatures).

Another consequence of mistakenly treating three checkerspot populations as one would have been the loss of information pertinent to an important theoretic question in ecology. Knowing how often populations go extinct in the normal course of events is very important to settling a major debate that has raged for years in the field of population dynamics. That debate has been whether population size in most species of organisms is regulated in a density-dependent or a density-independent manner.

If there is density-dependent population regulation, the size of the population in any generation is expected to be strongly influenced by the size of the previous generation. Suppose a checkerspot population gradually became very large. Each season the butterfly caterpillars (larvae) would eat more and more of the finite supply of food plants. One year, that supply might be exhausted, many of the larvae would starve, and the size of the population would decline. Eventually, regrowth of the food plant would again make it abundant relative to the needs of the reduced population of caterpillars. This would greatly enhance the survival of subsequent generations of caterpillars, and the population of butterflies would increase once again. The basic notion of density-dependence is that the abundance of the organism in one generation, in relation to its resources, determines its natality and mortality and thus whether the population size rises or falls in the next generation.

The key to density-dependent regulation is thus the size of the population relative to the carrying capacity of the environment. The carrying capacity is the maximum number of individuals that can be supported by a given habitat; it is usually related to the availability of a limiting resource, one that is in short supply in relation to the population's needs. Sometimes an absolute lid can be placed on population size by a nonrenewable resource such as available space. Only so many barnacles can be supported on a subtidal rock, for example. Holes and burrows (for animals that cannot dig their own) are also a limited type of space. Finger-sized saber-toothed blennies (reef fishes of the genus *Plagiotremus*) live in holes on the Australian Great Barrier reefs and dash from their shelter to slash bits off much larger fishes—and the number of blennies in an area is limited by the availability of appropriate-sized holes. But more often a limiting resource will be *renewable,* such as a population of mice preyed on by a population of owls, a population of larval food plants for a checkerspot demographic unit, or a population of a game animal for a group of human hunter-gatherers. Then the carrying capacity is determined by a maximum sustainable rate of exploitation. If that rate is exceeded, the resource is overexploited (the mice, food plants, or game are harvested more rapidly than they can be replaced), and the resource population declines, followed sooner or later by a decline in the population using that resource.

In density-independent population regulation, on the other hand, the availability of resources plays no significant role in changing the population size. The size of the population is kept below the carrying capacity by factors whose impact on the population do not differ much whether the population is large or small.

Suppose the number of checkerspots in a certain population was greatly reduced by unfavorable weather every third year on the average. These periodic reductions prevented the population from ever growing large enough to consume a major portion of its food supply. Changes in the weather could modify the supply of food plants for the caterpillars, but changes in the size of the checkerspot population would not.

Under these circumstances, one might expect that by pure chance several consecutive years of bad weather could cause a continual reduction of population size, leading to extinction. In contrast, extinction should be quite rare if regulation is density-dependent, because as the population shrank, improving conditions for each remaining individual ought to lead to higher survival and reproduction, and renewed growth of the population. Our first observation of an extinction suggested that Bay checkerspot demographic units on Jasper Ridge might be regulated in a density-independent fashion, as are many other plant-eating insects, including various pests of crops. Subsequent work has confirmed this impression. To explain what we know of the mechanisms of that regulation, I must first describe the life cycle of the Bay checkerspots more thoroughly.

Adult checkerspot butterflies are on the wing in March and April, the exact timing depending upon the weather. Clusters of eggs are laid on or around *Plantago erecta,* a small annual plantain related to the weedy European plantains that grow in many gardens and vacant lots. The young caterpillars hatch and are immediately engaged in a race against time. They must grow to a certain size before their food plants flower, produce the seeds of the next generation, and die as the annual California summer drought sets in. If they reach the right size, the partially grown caterpillars (less than a half inch long) then spend the summer dry season in diapause. They do not become active again until after the winter rains have begun and the next generation of their food plants has sprouted.

Through January, February, and early March, the revitalized caterpillars grow rapidly, and then form a chrysalis (pupa) in which the larva, essentially an eating-growing machine, is disassembled and the

An adult female Bay checkerspot butterfly laying a mass of eggs on plantain.

Caterpillar of the Bay checkerspot after diapause, feeding on plantain.

Chrysalis (pupa) of the Bay checkerspot in a depression in the soil. Tip of mechanical pencil at left for scale.

adult reproducing-dispersing machine is constructed from the parts. After a couple of weeks as a chrysalis, the adult emerges and mates, the female lays eggs, and the cycle begins all over.

A female checkerspot can lay more than 1,000 eggs, but obviously only a very small fraction of those can survive in most generations, or the entire world would soon be hip-deep in checkerspot butterflies. In fact, if the checkerspot population only doubled with each annual generation, in less than one hundred years the checkerspots would outweigh Earth itself. Such is the power of what the Reverend Thomas Malthus, the man whose name is forever associated with the concept of human overpopulation, called population increase "in a geometrical ratio." But the number of larvae that hatch from the eggs produced by an average female checkerspot (the natality of the population) varies little from generation to generation. Since variation in natality is small, it usually is not a major factor in the dynamics of populations of checkerspots.

Why, then, if natality varies little and the demographic units in areas C, G, and H all are subject to more or less the same climate, did one of the checkerspot populations on Jasper Ridge die out while its next-door neighbor population exploded? To answer that question, we must look for sources of variation in mortality that could balance the enormous reproductive potential of the butterflies.

The notion is widespread among ecologists that changes in death rates usually explain the dynamics of invertebrate populations, and our work on checkerspots supports that belief.* The relatively few early

* Mortality is not always the key to understanding population dynamics. This is especially true for mammals, whose reproductive rates may change quite dramatically with environmental conditions. But mortality can be extremely important in the dynamics of mammal populations, even looming as a major factor in future changes in the size of populations of *Homo sapiens*. Sadly, demographers (scientists who study human population dynamics) now always seem to focus on changes in natality when projecting future human population sizes. This is ironic, since demographers know very well that changes in mortality have played a major role in human population dynamics in the past. Humanity has been astonishingly successful over the past 10,000 years in first gradually and then rapidly reducing death rates—especially infant and child mortality rates—in its populations. And it was by lowering the output (the mortality rate) while leaving the input (the birth rate) largely unaltered that the population explosion was generated.

Now attention is concentrated, quite properly, on ways to reduce the birth rate and restore the balance. That makes sense, for humanity is the only species deaths we have observed among adults could not possibly account for significant fluctuations in population size, however. Thus death had to overtake most of the checkerspots between the egg and adult stages, but finding out when and why was likely to require a great deal of patient investigation in the cold, rain, and fog of the winter. I therefore decided to get a graduate student preadapted to those conditions to turn loose on the problem. Michael Singer, from England, proved the perfect choice. Mike disappeared into the field for a couple of seasons and returned somewhat damp, but with the answer.

It turns out that only a tiny fraction, perhaps 1 or 2 percent, of the larvae that hatch out of the eggs manage to grow large enough to enter diapause successfully before their food disappears. Most of the caterpillars starve, not because other caterpillars ate their food (as they might if population control were density-dependent), but because their food plants dried up in the late spring.

There seem to be three ways in which the race to enter diapause can be won by the caterpillars. Two are relatively unimportant, accounting for only a small fraction of the larvae that survive. If larvae are lucky enough to come from an early egg mass laid on plantain that happens to be growing in a relatively moist spot, the plant may last long enough for the little caterpillars to make it through. Or if the egg mass was laid on a plantain clump that is growing in soil "tilled" by a gopher, the larvae also have a chance, because the roots of the plantain can go deeper in the loosened soil, and the plant may not dry up as early as

---

that can consciously control its birth rate. In fact, one can argue that the most "human" behavior of *Homo sapiens*—the thing that most clearly separates our species from other animals—is the practice of birth control. And the social, political, economic, and technological questions surrounding that practice are among the most critical ones facing our species.

All of the emphasis on human birth rates, however, has distracted attention from the continuing potential for change in human death rates. Humanity has managed, primarily through the invention of agriculture, to increase the carrying capacity of the planet for people. Further, by exploitation (and destruction) of a one-time bonanza of resources—such as fossil fuels, concentrated mineral deposits, deep agricultural soils, extensive underground stocks of fresh water accumulated over hundreds of thousands of years, and diverse other organisms—humanity has increased its population size beyond a level that is permanently sustainable. This inevitably means that death rates will rise, a process that has already begun in some areas, such as the Sahel. Humanity will pay the price for exceeding the carrying capacity of its environment as surely as would a population of checkerspots.

others. Of course, to be successful with this gopher-plantain strategy, the larvae have to be lucky enough not to be on a plantain that is eaten by a gopher.

The vast majority that survive, however, do so for entirely different reasons. When the plantains on which most larvae are feeding do dry up, the hungry young larvae then start wandering. If they are fortunate enough to come across a plant of owl's clover *(Orthocarpus),* they can scramble onto that and complete their feeding. The owl's clover is an annual plant in the snapdragon family and happens to be one of the few plants other than plantain on which Bay checkerspot caterpillars can develop successfully (although Jasper Ridge females usually do not lay their eggs on it). Because owl's clover lives longer after the rain's end than does plantain, it plays a key role in the dynamics of the Bay checkerspot at Jasper Ridge.

In the early 1970s, we were able to show that the size of the checkerspot populations in one year was strongly correlated with the abundance of owl's clover available to the larvae in the preceding year. No density-dependence here—there were not enough caterpillars to reduce significantly the amount of owl's clover. But the more abundant the owl's clover, the better the chance that wandering checkerspot caterpillars would find some. We were even able to publish in *Science* magazine a prediction of 1975 checkerspot population size based on the density of owl's clover in 1974. Our prediction turned out to be correct—a relatively unusual occurrence in studies of the dynamics of natural populations, since the causes of changes in population size are rarely understood. But, as often happens in ecology, the solution to one problem begets another one. Now our group is trying to figure out what controls the population size of the owl's clover!

I relearned a big scientific lesson through Mike Singer's work— beware of things you take for granted. When I sent Mike out into the field, I told him that the checkerspot's larval food plant was the plantain. That's what I had been told by local lepidopterists, that's what the literature said, and I had successfully raised caterpillars both on that plantain and on its weedy relatives. The thought that there might be a secondary food plant had not even crossed my mind—indeed, at the time the concept of secondary food plants was essentially nonexistent among those who worked on butterflies. Mike's discovery of the owl's clover connection reminded me once again of a scientist's constant need to question the accepted.

Mike's discovery uncovered an important factor in the control of the dynamics of the Jasper Ridge checkerspots—the critical role of the owl's clover in determining patterns of mortality. When owl's clover is abundant in an area, that will increase the chances that the checkerspot population there will expand; when owl's clover is scarce, the butterfly population is likely to contract. But there is more to the story than that, and our group is still working hard to unravel the details. It is now clear, for instance, that differences in microclimate near the soil surface where the caterpillars live can play a major role in the dynamics of the checkerspots. Although areas C, G, and H are exposed to the same climate, each contains a different set of microclimatic regimes because of different degrees of slope and exposure. A year that may be too cold and wet for ideal caterpillar growth in a north-facing swale may be just right for larvae living on a south-facing slope. But even microclimate and the abundance of owl's clover (itself undoubtedly influenced by the microclimate) do not seem to be the entire story—especially of the extinction of the demographic unit in G—as we'll see in a bit.

Early in our group's work on the Bay checkerspot, I realized that if we concentrated just on the Jasper Ridge populations, we could learn a tremendous amount about them, but we would have no idea how broadly applicable our conclusions might be. Were the Jasper Ridge populations typical of all those of the Bay checkerspot, or of all Edith's checkerspot populations? (Remember, the Bay checkerspot is just one subspecies of Edith's, *E. editha bayensis*.) Are they typical of herbivorous insects, or, in some respects, of all animals? To answer such questions, we branched out and began to investigate nearby populations of the Bay checkerspot, of other subspecies of Edith's checkerspot, of other closely related checkerspot species (especially the chalcedon checkerspot, which also occurs on Jasper Ridge), and of totally unrelated butterflies. Mike Singer, and Larry Gilbert, another of my graduate students, did pioneering work with other populations of Edith's checkerspot within California, populations which our group has continued to investigate for the past fifteen years or so. This work has demonstrated that species are not necessarily ecologically uniform. Studies of Edith's checkerspot populations throughout the western United States have shown that the ecology of populations of the Bay subspecies is not typical of the entire species. In different areas, the species attacks different food plants, the adults fly at different times, and their populations have different structures. While many populations of Edith's checkerspots, like those on Jasper Ridge, are controlled in a density-independent manner, others show density-dependent control.

These findings should alert those charged with combating economic pests such as the Mediterranean fruit fly (Medfly) that they should plan their control measures to deal with the ecology of the pest populations in specific locations, not around the presumed attributes of a widespread species. Many pests show variation in ecological attributes similar to those in Edith's checkerspots—indeed such variation is probably the rule rather than the exception.

Not only did the studies of other checkerspot species reveal a diversity of ecological patterns, so did investigations of butterflies in entirely different groups. Another first-rate student of mine, Pete Brussard, and I examined the population structure of a widespread wood nymph butterfly of Colorado mountain meadows. We found that its demographic units covered areas orders of magnitude larger than those found in the Jasper Ridge checkerspots.

But what interested me most was whether butterflies in a tropical rain forest would show the same patterns of fluctuations and extinctions as the demographic units of Bay checkerspots on Jasper Ridge. The rain forests are arguably the most important habitats on Earth—nature's great storehouse of diversity. They are rapidly disappearing and the biology of their inhabitants is little understood. The folklore of ecology held that butterflies (as well as other organisms) in the jungle had much more stable population sizes than their counterparts in the temperate zones.

To see if that were true, in the early 1970s, Larry Gilbert and I started a long-term project on the dynamics of a longwing butterfly, *Heliconius ethilla,* in the mountains of northern Trinidad. Our group worked out of the New York Zoological Society's field station at Simla, which had been established decades earlier by the famous naturalist William Beebe.

It was a most pleasant base of operations. Following a long, hot day of marking and recapturing butterflies in the mountain forests of Trinidad, we would return to a fine old shaded house, shower, and have cocktails. After dinner we could work up our data, consult the station's excellent library, or marvel at the incredible diversity of moths and other insects attracted to the "black light" (which produced ultraviolet light, visible to insects, as well as light of wavelengths visible to us) that we kept on the back porch. And, if we were feeling ambitious, Mike Singer and I would sometimes go out with head lamps and crawl around in the undergrowth looking for the nocturnal caterpillars of *Euptychia* butterflies, whose interactions with their food plants (grasses and their relatives) the group was also studying. Of course, on those forays, we had to keep a very sharp eye open for bushmasters. Nocturnal relatives of rattlesnakes, bushmasters are the largest and deadliest of pit vipers—and the most impressive local hazard.

Our work at Simla was cut short by the unexpected closing of the station, but not before we had made some very interesting discoveries. For one thing, the longwing populations showed very little fluctuation in comparison to Bay checkerspot populations. In this case, the folk wisdom about the tropics was supported by evidence. The longwing butterflies also proved to be very long-lived. Our recapture records showed them living up to four months as adults (in contrast with an average of perhaps seven days for a checkerspot). Longwing adults sometimes lived long enough to have adult grandchildren flying with them, a very unusual life-span for an insect.

In fact, the longwings have a totally different life-history strategy from the checkerspots. Checkerspot females emerge from their chrysalises with a large supply of mature eggs, and they dump them as quickly as possible in order to give their larvae the best possible chance of reaching diapause before the food plant is no longer available. Longwing females, in contrast, emerge with no mature eggs and lay just a few eggs a day throughout their long lives. It is easy to speculate on the evolutionary reason for the different strategies, but virtually impossible to demonstrate whether the speculation is correct.

One speculation involves ants, the most common and ferocious predators in the tropical-forest habitat of longwings. Ants scour the vegetation, including the passion-fruit vines that provide food for the longwing larvae, looking for caterpillars and other prey. Thus the life of a longwing caterpillar in a tropical forest is a very precarious one. Since females emerge with no mature eggs, the larval stage need last only long enough to put aside reserves for building the adult structure; no extra resources need be stored in that stage for egg production. Thus, evolution may have tackled the problem of ant predation by minimizing the amount of time that longwings must spend as caterpillars. The female longwings have the task of acquiring the resources they need for egg production, since being an adult longwing is a lot safer than being a larva. (An additional longwing evolutionary strategy is that their caterpillars extract poisons from the passion vines, which are passed on to the adult stage. These make the adults distasteful to their main enemies, birds, which appear to be more fastidious than ants.)

But where do the adult female longwings get the nourishment required for egg production? Adult butterflies sip nectar, which is mostly sugar; proteins, however, are needed to produce eggs. It was Larry who solved the puzzle, at the same time reminding me again of the dangers of intellectual rigidity. He noted that longwings often have a lot of pollen on their long, coiled "tongues." I too had noted large pollen loads many times, but had always assumed that it simply showed that the flower had successfully attracted a pollinator. After all, the reason that plants produce flowers and provide nectar for butterflies, bees, and other insects, and certain birds, is to lure them into helping the plant with its sexual needs. In the process of getting the nectar, the butterfly transports pollen.

So I "knew" what was going on. But Larry was more flexible in his thinking and a better observer. He noted that the longwings seemed to be manipulating the pollen with their tongues. In a series of experiments, he was able to show that they were, in fact, digesting it outside of their bodies with regurgitated saliva and then imbibing the rich broth of amino acids—the building blocks of proteins. His discovery was especially embarrassing for me because, in my dissertation studies of butterfly morphology, I had noted the tiny pegs on the tongue that the longwings use for manipulating the pollen into little balls. But I had totally missed their function, assuming them to be sensory organs.

To obtain the pollen they need, the longwings patrol "traplines"— visiting in a regular sequence the pollen-producing flowers of certain rather widely dispersed jungle vines; vines that are an indispensable resource for the population. As Larry has subsequently shown, these butterflies are not only long-lived, they also are more intelligent than their temperate-zone relatives on Jasper Ridge. The females remember the locations of pollen sources, and they also repeatedly examine the same passion vine, waiting for precisely the right stage of leaf development before laying their eggs.

The reproductive strategy of the checkerspots is totally different. Nectar and pollen resources are even more short-lived than plantain leaves in the California spring. Checkerspot larvae also do not live in an environment that is swarming with ferocious ants, and, as far as we can tell, are subject to much less predation than the larvae of longwings.

One would therefore expect that the need for adult feeding has been reduced to a minimum. There would be little to gain by shortening a female's larval life so that she could emerge and attempt to acquire resources for egg production as an adult. So checkerspots have evolved to spend most of their lives as larvae, eating enough to provide construction materials both for making adults and for producing of a lot of eggs.

After Larry discovered the significance of pollen-gathering in longwings, I began to wonder why checkerspots drank nectar at all. Some checkerspot populations seem to thrive with little or no available nectar. And yet, when appropriate flowers are abundant, adults spend a great deal of time sipping from them. Standard evolutionary thought holds that organisms are very unlikely to put much effort into an activity that doesn't increase their reproductive potential. I therefore arranged for one of my graduate students, Dennis Murphy, to pursue a most difficult dissertation project—to see if he could discover how the nectar resource influenced the dynamics of the butterfly populations.

I knew that even if he failed to answer that question, at least Dennis would learn a lot about the foraging strategies and use of nectar by checkerspots. (There was little chance a dissertation would not come from his efforts; a crucial consideration, needless to say, when planning a research project with a doctoral student.) But Dennis did a superb job and came up with an answer. From feeding experiments, he was able to show that the first two egg masses laid by a female Bay checkerspot on Jasper Ridge are about the same size and weight whether or not the female has fed on nectar. And those are the two most critical egg masses, since usually they are the only ones producing larvae with a chance to reach diapause.

Nectar *was* very important, however, in maintaining egg production after the first two masses were laid. Sipping nectar can greatly increase a female's egg production after the first two egg masses; it can also keep a male going long enough to fertilize some of those additional eggs. This is not significant in most years because the early senescence of the food plant means death from starvation for most larvae hatching from late egg masses. But, when the weather is especially favorable, the plants last longer, and larvae from later egg masses can survive, thus greatly increasing the size of the population.

Dennis suggested that, at Jasper Ridge, the demographic units of Edith's checkerspots that had the most suitable nectar sources would be the best buffered against extinction—because, in a population whose fluctuations are density-independent, large population size provides insurance against accidental extinction. This hypothesis would explain the eager nectaring by checkerspots every season, since they have no way of predicting the time of plant senescence. It would also explain the extinctions in area G, since larval food has always been abundant there, but the nectar resources have been relatively scarce. If Dennis's hypothesis proves correct, the demographic unit in area G has simply not been able to build populations large enough to provide reasonable insurance against accidental extinction.

Now that the dynamics of populations have been introduced, we can turn to a consideration of their evolution. The two topics are intertwined. Natality and mortality are not of interest only in connection with the dynamics of populations; those rates are critical to understanding the evolution of populations as well, and birth and death rates are often modified in the course of evolution. The great reproductive capacity of organisms was known to Victorians through the writings of Reverend Malthus and did not escape the attention of Charles Darwin, who first made the connection between the dynamics and the evolution of populations. Darwin realized that the vast majority of offspring produced by most organisms failed to survive to reproduce. What, then, he asked, determines who succeeds in reproducing and who does not? In an analogy to the practices of animal breeders and plant propagators, he surmised that nature selects the offspring that are going to be the parents of the next generation. And which ones does nature select? Those that are best able to survive and reproduce.

This somewhat tautological concept of natural selection—that the fittest survive and those that survive are the fittest—is, even today, the centerpiece of evolutionary theory. Nature has no goals. A breeder of cattle may select for breeding the cows that produce the most milk; a geneticist working for an agribusiness company may select as parents of the next generation the tomato plants that produce the most cubical, most durable, and (coincidentally) most tasteless tomatoes. But reproduction is an end in itself in nature, and those that are best at it are, by definition, the most fit. Natural selection explains why organisms are so well suited to their ecological contexts. Only by invoking selection can we comprehend *why* nature's machinery functions as it does.

In modern terms, natural selection is occurring whenever there is differential reproduction of genetic types. If individuals with one kind of genetic endowment regularly outproduce those with another, then natural selection is occurring. Fitness, in evolutionary theory, is a measure of the reproductive capacity of one genetic type in comparison with another. Fortunately, for reasons we'll explore in a bit, all individuals are not genetically identical. If they were, there could be no differential reproduction of genetic types, and no evolution.

In the shorthand of geneticists, a genetic type is called a genotype. This is differentiated from a phenotype, which is simply the physical appearance and functioning of an individual organism. The phenotype is the result of the interaction between the genotype and its environment. For example, children have an inherited tendency to grow to a certain height, but how tall they actually grow will depend on both that genetic tendency (heredity) and on environmental factors such as diet. A man who has "tall genes" but is malnourished as a child may end up just over five feet tall—the interaction between his tall genotype and a poor environment may produce a short phenotype.

The most fit genotype in a population is, by definition, the one that on the average produces the most offspring in a given generation. Fitness in this technical sense, sometimes called Darwinian fitness, is a quite different thing from fitness as it is used in everyday language. It boils down to the proportional contribution of an individual's genes to posterity—that is, how well it reproduces relative to other individuals in the population. A six foot five inch, supremely healthy, sublimely handsome, childless twenty-two-year-old man who has had a vasectomy is not usually as fit in the Darwinian sense as a misshapen, diseased man who has sired a healthy child.

I added the caveat "not usually" because fitness is somewhat more difficult to determine in social organisms like human beings than in nonsocial organisms. The name of the game is not just reproduction, but reproduction of genes like yours. In a social organism, a sterile individual may promote copies of its genes by, for example, helping out its siblings (each of which has half of its genes identical to those of the sterile individual). Thus our sterilized man, if he had saved five brothers and four sisters from a fire, may actually have been, in an evolutionary sense, much more fit than the unfortunate deformed father. Later, when social behavior is discussed, you will see the importance of this expanded view of fitness.

In the present context, however, it is important simply to remember that differential reproduction of genotypes is the basic driving force of evolution. It explains the slender, streamlined form of the fish, the intricate marvels of the human eye, and the dazzling colors of a morpho butterfly's wing. It also explains why checkerspot butterflies emerge from their chrysalids with many eggs and longwings with few, and it explains most of the features of the ecological relationships and systems that are described in this book. The great triumph of mathematical population genetics has been to show that even very slight reproductive differentials—one type, say, producing 101 offspring on the average while another produces only 100—are quite sufficient, given the enormous amount of time available, to explain the course of evolution. Such slight differences can account for the evolution of life from simple protoorganisms, barely differentiated from the thick organic soup of some early marine environment, all the way to the upper-class Englishmen who, in Darwin's day, were thought to represent the pinnacle of evolutionary progress.

My first conscious contact with selection came in the Animal House. This was not the fraternity of movie fame, but rather an old building on the University of Kansas campus in which Robert Sokal (a brilliant young assistant professor who in a few years would revolutionize the science of taxonomy) was carrying out experiments on the evolution of DDT resistance in fruit flies. I had gone to the University of Kansas to do graduate work under a distinguished evolutionist, Charles Michener, but I was supported by a research assistantship under Sokal.

The goal of Bob's project was to discover whether the evolution of resistance to DDT in fruit flies also changed their behavior. Fortunately for us (but unfortunately for pest and disease control), producing resistant flies was ridiculously easy. All one had to do was add DDT to the gooey nutritional medium on which the fly's eggs were placed. The maggots developed on the medium while feeding on fungi that also grew there. Enough DDT was added to the medium to kill most of the flies in each generation, and then the survivors were used as the parents of the next generation. After ten generations or so, this regime produced a strain of fruit fly that enjoyed a slug of DDT as an apéritif.

Producing strains of fruit flies that were more susceptible than usual to DDT was more difficult, however. Obviously, it was impossible to use as parents of the next generation the flies that died most easily when exposed to DDT in their larval stage. This problem was surmounted with a technique known as sib selection. Eggs from each pair of flies were divided into two batches. One batch was placed in a vial containing the medium laced with DDT. The other batch was placed in a vial with unpoisoned medium. The vials were carefully cross-numbered. When the next generation of flies hatched out, we could see which of the DDT-laced vials had the highest mortality. We then went to the DDT-free vials and used as parents of the next generation the brothers and sisters of the flies that had died in the greatest numbers. Using this procedure, in some ten generations we produced strains in which the flies just about dropped dead at the sight of a bottle of DDT.

These techniques, of course, were old hat to geneticists of the early 1950s, but as a beginning graduate student I found them fascinating. They showed me how easily selection with a goal determined by human beings—artificial selection—can shape a population. And that greatly deepened my interest in the processes of evolution, complementing an already strong commitment to the study of its products—the vast diversity of living things. In addition, the experiments I did with Sokal impressed me with how important natural selection was even in the laboratory. It was clear from decades of experiments by many investigators with fruit flies and other laboratory organisms that the selection imposed by the investigator was virtually always accompanied by natural selection. For example, if we stopped selecting the parents of each generation for resistance or susceptibility to DDT (that is, if each generation was no longer exposed to the poison), the characteristics of either the resistant or the susceptible strains tended over succeeding generations to return to those of control strains that had never been exposed to DDT. Clearly, natural selection was opposing the artificial selection that we applied.

As I mentioned earlier, selection could not operate if all organisms were identical genetically. There must be genetic variation for there to be evolution. But what mechanisms produce the required variation? What keeps all genotypes from being the same? One mechanism is inherent in the system that passes genetic messages (basically, sets of instructions for making a new organism) from parents to offspring. When organisms reproduce, their offspring usually do not receive precisely the same set of messages that the parents had. One reason is that external agencies, such as ionizing radiation and certain kinds of chemicals, can cause alterations in the genetic messages. Another is that mistakes are sometimes made when a molecule of deoxyribonucleic acid (DNA), which contains the coded genetic instructions of the parent, is copied. DNA is copied each time a cell divides and whenever the organism reproduces. The changes, whether caused by external agencies such as radiation or internal mistakes, are called mutations, and these are one source of the crucial genetic variability.

Sexual reproduction adds another dimension to genetic variability. Sex is a mechanism by which new combinations are made of genetic variants. The process is a little complex, but some understanding of it will be helpful to you when we deal with the evolution and ecology of sex. The reshuffling, or, more technically, the recombination, of genetic variants occurs during a complex series of changes that occur in the chromosomes, the tiny rod- or threadlike structures that contain the DNA of the cells (and thus the genes, which can be thought of as units of heredity—individual messages within the multitude of instructions coded into the DNA). Organisms may have dozens of chromosomes in each cell (people have forty-six), each of which contains thousands of genes.

Basically, except for very specialized cells, the cells of the vast majority of higher organisms contain two sets of chromosomes; they are said to be diploid. One set of chromosomes traces to the mother, and the other to the father. When such a cell divides (during normal growth or to repair an injury), the DNA in the chromosomes is copied, and each chromosome splits longitudinally into two. One daughter chromosome of each pair goes to each of two daughter cells. This process, known as mitosis, guarantees that each new cell is also diploid; it receives two complete sets of chromosomes and thus a complete set of genetic instructions.

Sexual reproduction, however, is quite different, since two cells (sperm and egg) fuse, combining their sets of chromosomes. Without some compensating reduction in chromosome number, each generation would have twice as many chromosomes as the previous one. What intervenes is a special process of division occurring in the precursors of the sperm and egg cells that halves the number of chromosomes, so that each sperm and egg cell has just one member of each chromosome pair. That process, called meiosis, changes the diploid number of chromosomes, cutting it in half to what is called the haploid number. Then when the haploid sperm and egg fuse (at conception), the diploid number is restored.

Convenient as the "reduction division" of meiosis is for avoiding a buildup in the number of chromosomes, that is not its only function. Another important feature of meiosis is a maneuvering together of the pairs of chromosomes, one from each parental individual, during which members of pairs can exchange segments of chromosome. Thus genes can be passed from one member of a pair of chromosomes to another, which could not happen without meiosis. This exchange of genetic material between chromosomes that came from the father and chromosomes that came from the mother is the physical basis of recombination. As a result of such exchanges, each egg or sperm you produce contains a mixture of genes from your two parents, even though each egg or sperm carries only a single set of chromosomes. While mutation is the most important long-term source of variability, recombination is the most important immediate source of the genetic variability that permits natural selection to operate.

Let's look at how sexual reproduction increases the variability that selection has to work with. Suppose, for the moment, that human beings reproduced asexually, as many organisms do. Barring mutation, a mother who had a genotype making her blue-eyed, color-blind, and of blood-type B would produce nothing but blue-eyed, color-blind, type B daughters. Another mother who had brown eyes, normal vision, and blood-type A would produce only daughters with that phenotype. Now suppose that the human population were made up exclusively of females of those two types. Then only one pattern of differential reproduction would be possible. One type would outbreed the other and, presumably, replace it eventually—unless mutation intervened. Mutation could produce, say, blue-eyed, normally sighted, type A individuals who might be superior to either or both of the other types. But mutations are very rare events—a given gene normally will mutate only on the order of once in every 10,000 or more individuals that carry it. Therefore, especially in a relatively small population, the waiting time for mutation to create many of the possible genetic combinations of eye color, color blindness, and blood type would be hundreds or thousands of generations.

On the other hand, if half of the females in our hypothetical population were replaced with males—in other words, if sexual reproduction were the rule—all possible combinations of the three traits, as well as some novel types such as individuals with blood-type AB, would show up in the first two generations. And all of the combinations might have different reproductive capacities. The example is oversimplified, but the principle is important. Everything else being equal, a sexually reproducing population will contain a much more varied array of genotypes than an asexually reproducing population.

Another important point to be made here is that, contrary to some religious dogma, the "purpose" of sex is *not* reproduction. As the life histories of many organisms show, reproduction is a great deal easier without sex. If reproduction is asexual, there are no problems of finding mates, and populations can produce a lot more individuals since all individuals are female and devote all of their energy to reproduction, rather than having to convert some of the population's resources into producing "useless" males that do not bring forth offspring. One important function of sex in all organisms is recombination, but in many animals, including human beings, sex also has a variety of additional social functions. Exactly why sex evolved remains a central mystery of evolutionary theory—since it is not crystal clear that the additional variability provided by recombination is a net evolutionary benefit, as we'll see in the next chapter.

One can think of mutation and recombination as processes producing genetic variation, and of selection as a process for sorting it out, generally reducing the amount of variation by discarding traits that do not help the individual cope with its environment. But how much genetic variability is actually present in natural populations has been a topic of continuing controversy—a controversy very important to understanding how evolution works.

Population geneticists have held two opposing views on genetic variability. One school, tracing in part to the great geneticist H. J. Muller (who first showed that radiation could cause mutations and found that most of those produced were deleterious), has held that populations generally consist of individuals with similar, standardized genotypes— the "wild type" (so called originally because it was the type found in "the wild" rather than in the laboratory). This school views mutations as causing deviations away from this wild type and, for the most part, as being bad for the organism. The idea is that a random change is unlikely to improve a plant, animal, or microbe that is already functioning satisfactorily in nature.

These geneticists see a major role of natural selection as purging deleterious mutants from populations, and they would expect the genetic basis for most characteristics in the population to show little variation. Note that they do not deny the efficacy of natural selection in producing the sharp teeth of sharks, the alluring (to insects and other pollinators) characteristics of flowers, and most of the other features of organisms. They just think that, in most populations most of the time, the majority of individuals share overwhelmingly the same genetic code.

They think variability is relatively scarce and is created primarily by a balance between the production of mutations and the purging of mutations by selection.

A somewhat different view has been held by a school associated with one of the greatest evolutionists of the twentieth century, Theodosius Dobzhansky, who spent much of his career studying variations in the structure of the chromosomes of fruit flies. This school expects selection to maintain a great deal of genetic variation in populations, because its adherents think that possessing genetic variation itself provides an evolutionary advantage. They believe the advantage takes several forms. One is that having different kinds of individuals in the population allows members of the same population to fit into slightly different ecological roles, permitting a larger population size. Another is that the variation is useful because it permits the rapid evolution necessary for the survival of the population when the environment changes.

Until the late 1960s, the argument between the Muller and Dobzhansky schools raged pretty much in a vacuum, because there was no convenient way of measuring the amount of genetic variability in a large sample of populations. In the late 1960s, however, it looked as though the problem would soon be solved because a technique for measuring the amount of genetic variation in proteins became available. That technique, called electrophoresis, is described in Appendix A.

Using electrophoresis, biologists can detect genetically caused differences in the proteins of virtually any organism, and this has produced an extraordinary opportunity to assay the genetic variability of populations. Most populations that have been studied with this technique show a high degree of genetic polymorphism—the presence of individuals that differ in the kinds of genetic information they carry (such as two genes producing two different eye colors or two different forms of a particular protein). Thus, by around 1970, the view of the Dobzhansky school appeared to be supported. There was a lot of genetic variation in natural populations—more than could be accounted for if they were mostly composed of wild-type individuals and selection were largely "purifying"; that is, removing deleterious mutants. Selection had to be favoring the polymorphisms.

Or was it? The Muller school counterattacked. Under the lead of Japanese geneticist Motoo Kimura (Muller had died), a new proposition was put forward: *most* of the variation exposed in the protein studies was of no biological significance. The proteins investigated were primarily enzymes—the biological catalysts that play critical roles in the functioning of all organisms. This "neutralist" school claimed that, in most cases, the small changes in the enzymes revealed by electrophoresis made little or no difference to the functioning of the enzymes. Selection therefore could not operate at all on them; there could be no differential reproduction of the different genotypes because the organisms possessing them were functionally the same. So there *was* a lot of genetic variation, but it didn't mean anything. The genetically different individuals all produced the same wild phenotype, and selection's main role was still removing the relatively few mutations that changed enzymes enough to affect significantly the organism's phenotype.

As one might expect, the "selectionists" retorted that it wasn't so. They claimed that the vast majority of changes in enzymes did affect the functioning of the enzymes sufficiently to change phenotypes and permit selection to work. Both neutralists and selectionists have carried out a long series of studies attempting to determine which of the two views is correct.

Finding the answer is very important to both ecologists and evolutionary biologists. For instance, one of my main goals when I started the work on checkerspot butterflies was to look at their ecology and evolution simultaneously, to see, for example, how changes in the dynamics of populations affected their genetics, and vice versa. The electrophoretic techniques seemed, at long last, to open the door to studying a reasonable sample of the genetic endowment of the butterflies. Perhaps we could answer questions such as "When the population size declines, is genetic variability greatly reduced?" Thus, as soon as we adapted the electrophoretic techniques to work on butterflies (it was a fairly long struggle), our group began to look at enzyme variation in Edith's checkerspot.

Then along came the neutralists to claim that the variation we were studying was of no significance. Like it or not, we needed to have a reasonable answer to the neutrality controversy before we could make a sound interpretation of our data. Unfortunately, despite the valiant efforts of many population biologists, there has not been a satisfactory resolution to the problem. If you would like to know a little more about how it has been approached, I've given some details in Appendix B.

Meanwhile, the lack of a resolution has meant that our research group has been faced with a rather common scientific dilemma. How much effort should be expended on an important but apparently intractable problem? When should you stop throwing good money after bad; when should you cut your losses and redirect your research? The neutrality controversy is central to some of the most interesting questions in population biology, but it looks as though some sort of conceptual breakthrough will be required for it to be resolved satisfactorily within the lifetime of our research project. But the hope for a breakthrough is always there.

For example, recent extinctions of several *Euphydryas editha* populations during the California drought of the mid-1970s gave us an opportunity to make a unique test of how well the neutrality hypothesis holds for checkerspots. We raised butterflies in the laboratory and used them to re-establish populations in areas where they have gone extinct. The colonists we are using have frequencies of enzyme variants very different from those that originally existed in the now extinct populations. If the recolonization experiments work and new populations are established, we will monitor what happens to those frequencies. If they change toward the previous frequencies over several generations, that will be powerful evidence that selection is at work. It would mean that environmental conditions at the population site favor a certain frequency of enzyme variants and that selection can detect the differences between the enzymes. If they do not change in that direction, the neutrality hypothesis will be supported.

Unfortunately, such experiments take a long time to run—we always wait at least two years to be sure that a population is extinct, and then we must wait several more years after recolonization is attempted before the population is large enough to allow us to remove an adequate sample of individuals for genetic analysis. The work is also very laborious. Before recolonization, caterpillars from many matings must be raised to diapause in the laboratory. Then colonists with the proper frequencies of enzyme variants are chosen from them (on the basis of the frequencies in the caterpillar's parents, which have undergone electrophoresis). But we have managed to start experiments at two different sites where populations had gone extinct, and it looks as if one may be concluded successfully; the population has been re-established. Of course, with our luck, the results may be ambiguous—say, the frequency of types of one enzyme will return toward the level of the previous population, and the others will fluctuate randomly.

A difficulty with the neutrality question is one that is frequently encountered in ecology: finding answers in a single system (species, community, ecosystem) is not enough. The question isn't whether things happen one way or another, but *how often* they happen each way. How frequently and strongly does selection operate on enzyme variants? What proportion of populations is controlled in a density-dependent manner? Ecologists therefore are often faced with serious questions about their sampling of nature. For example, population genetic theory has shown that relatively tiny amounts of selection occurring through the enormity of geological time can account for all observed evolution. But this does not tell us whether selection pressures are generally weak and constant, or whether they might, for instance, be very strong but intermittent. We need to discover how selection is operating in a wide variety of organisms, so we can see how often it is strong and how often weak.

Consider what is known about selection pressures in natural populations. In only a relative handful of cases has this been studied directly, the most famous example being that of the peppered moth, *Biston betularia,* in England. In the early part of the last century, virtually all individuals of this species of moth were "peppered"; that is, they had a mottled pattern that camouflaged them when they rested on the lichen-covered trunks of trees. But by the end of the century, peppered-moth populations in much of England consisted largely of very dark (melanic) individuals.

This change had occurred gradually and was associated with the industrialization of the country. The melanic populations were found in those areas where pollution from coal burning had killed the lichens and blackened the tree trunks. In a series of elegant experiments, the late Bernard Kettlewell, a British physician and biologist, showed that natural selection had caused the change.

The most dramatic experiments involved placing moths of both types on trees in polluted and unpolluted areas and photographing what happened. The resultant movies show natural selection in action; watching them is a thrill for anyone interested in evolution. On a blackened trunk, birds fly into view and work their way up and down the trunk, devouring the mottled moths and missing the melanics. On lichen-covered trunks, the opposite occurs; the birds miss the mottled moths and eat the melanic ones. Differential predation, leading to differential survival (and thus differential reproduction) of the two types of moths, was shown clearly to be a major factor in the development of what has become known as industrial melanism. Strong selection pressure was demonstrated in the field.

Thirty years ago, when I was a postdoctoral student with Joseph Camin at the Chicago Academy of Sciences, I became involved in another of the few field studies ever made of selection. Although at the time we were working on the transmission by snake mites of a malaria-like disease of snakes, our attention was drawn to an unrelated puzzle. Part of our project involved collecting common water snakes from an island in Lake Erie to use as experimental animals in the disease study. Most populations of water snakes are made up entirely of striped individuals, but the island snake populations had a high proportion of unstriped snakes.

When we took a close look at the situation, we discovered that there was a much higher proportion of striped individuals among newborns than among the adults. But the striped young stood out quite prominently against the flat gray limestone rocks of the island's shore; unstriped ones were harder to see. This and other evidence convinced us that visual predators (those that hunt by sight), possibly sea gulls, were attacking striped young snakes more often than unstriped ones. Continuing migration of striped snakes from the swamps on the lakeshore (where stripes helped camouflage the snakes) kept the Lake Erie island populations from becoming all unstriped. Strong selection and migration, acting in opposite directions, created a situation in which evolution could be seen in action.

In another well-known case, land snails with different patterns and colors on their shells were known to be associated with different habitats —for example, more snails with striped shells are found in rough herbage than in short turf, where bandless shells predominate. The demonstration that differential predation was responsible for this association was facilitated by the habits of thrushes that ate the snails. The birds would crack the shells of their prey on favorite rocks, which were called "thrush anvils." Around the anvils accumulated a sample of shells of the individuals the birds were finding and eating. By comparing the proportions of different types at the anvils with those in the snail population as a whole, two British scientists, A. J. Cain and P. M. Sheppard, showed that the snails taken by the predators were not random samples from the population. For instance, birds hunting snails in one habitat were more likely to find and eat banded than unbanded individuals, and, as a result, unbanded individuals predominated in that snail colony. Again, strong selection was found to be occurring.

The question remains: How typical are the strong selection pressures uncovered in these and the handful of other studies? From the studies that have been done, one might conclude that evolution has been zig-zagging along under the influence of strong selection tracking an ever-changing environment. In other words, strong selection operating in different directions at different times could, over long periods, produce products of evolution that (especially in groups without fossil records) are indistinguishable from those produced by constant weak selection operating in one direction.

But the question cannot yet be answered because the sample of studies of selection in nature is clearly biased toward those in which the operation of strong selection has attracted the attention of investigators. To understand nature as a whole, we need to avoid such bias. Indeed, our group's work with checkerspots was planned explicitly to be an element in a broader sampling. The checkerspots on Jasper Ridge were doing nothing to attract an investigator's interest when my studies began, and that was one reason I wanted to study them.*

So far I have discussed only examples of evolutionary changes *within* populations: populations of fruit flies becoming resistant to pesticides, or the peppered-moth population of England's industrial Midlands converting from mostly mottled individuals to mostly melanic individuals. This is, of course, a very important aspect of evolution. But to ecologists there is another equally important aspect—the diversifying process of evolution, the differentiation of populations into new species, or what is generally known as speciation. After all, if there were no means by which the tree of life could branch, the whole world might simply be covered with one highly evolved living slime. Instead, ecologists today are confronted with an extremely varied array of species, perhaps as many as 30 million different kinds of organisms. What can explain this enormous diversity?

The fundamental answer can be found in the diversity of physical conditions on the planet. What's sauce for a butterfly is not sauce for the gooseneck barnacle. It takes a very different kind of organism to fly through the air than it does to cling to a rock in the sea. Different

---

* A sampling problem exists in population dynamics as well as population genetics. Much more research has been done on large populations than on small populations. This is quite natural, since both pests, whose populations we wish to suppress, and economically valuable organisms (such as commercially harvested fishes), which we wish to exploit, tend to have large populations. But most natural populations are small, and their dynamics may be quite different from those of large populations.

physical environments impose different selection pressures. And just as we can see evolution changing a single population, so we can see evolution causing two populations to become increasingly different.

For example, the evolution of industrial melanism differentiated those populations subjected to a highly polluted environment from others that were not; the peppered moths in the Midlands became quite different from the peppered moths in other, relatively unpolluted parts of Britain. And most evolutionists believe that two populations in different areas responding to different environmental forces, as the peppered moths did, eventually may lead them to evolve into two separate species.

And how do we know when we have two species rather than two different populations of the same species? It is generally agreed that two differentiating populations, A and B, qualify as separate species when individuals in population A can no longer successfully interbreed with individuals of the opposite sex in population B. But how does the inability to interbreed arise?

Briefly, the answer is that varying physical environments subject populations *isolated from one another in different areas* to different selection pressures, which in turn leads to their becoming increasingly different genetically. As organisms in different areas become less and less alike, the communities of which they are a part automatically become more and more different. Differences in the biological communities then feed back on the speciation process by making the environments even more distinct (remember, all the evolving organisms of a community are part of one another's environments).

Eventually, the separated populations become so different that they are no longer capable of interbreeding and must go their own evolutionary ways, even if they subsequently come to occupy the same area again. Any of several different factors may be responsible for the loss of ability to interbreed. For instance, the signals used in courtship may have changed so that members of the differentiated populations no longer respond to each other's signals. Or the entire genetic codes of the two populations may have become so different that hybrid individuals are sterile (as are mules) or simply not viable (the hybrids die before they mature).

The emphasis on geographic separation as a requirement for speciation is an old one. Since the publication of Darwin's *Origin of Species,* the differentiation of an ancestral South American bird into the array of species of Darwin's finches on the various islands of the Galapagos has

Left, Darwin's finches on the Galapagos Islands are a group of species that have differentiated largely on the basis of the size and shape of their beaks and their diets. This species of ground finch has a relatively large bill and feeds on moderately hard seeds and some insects.

Right, this species of ground finch has a smaller bill than the one in the other photo and presumably feeds on softer, smaller seeds and some insects. Darwin's finches include species with relatively slender bills, as well as sparrowlike bills as shown here. The woodpecker finch, with a tanagerlike bill, occupies a niche like that of a woodpecker (of which there are none in the Galapagos). It has not, however, evolved the long tongue that woodpeckers use to extract insects from the holes they drill. Instead the finch often uses a tool—a cactus spine—to extract its prey.

been considered a classic case of speciation in isolation. In 1904 the great American ornithologist Joseph Grinnell stated the principle explicitly: "It is *isolation,* either by barriers or by sufficient distance to more than counterbalance inheritance from the opposite type, that seems to me to be the absolutely essential condition for the differentiation of two species, at least in birds."

The process of speciation is much more difficult to observe than is simple evolution within populations, because it normally takes a much longer time. Evolutionists think in terms of tens or hundreds of thousands of generations for a single species to divide into two or more daughter species (although things may go much more rapidly or much more slowly, depending on a variety of factors). Consequently, the best evidence that speciation occurs more or less as described does not come from actual observations of species A turning into species B and C, but rather from the state of organic diversity at the present time.

Ecology is the key to speciation. Evidence abounds that different environmental situations cause populations in different geographic areas to evolve diverse characteristics. Geographic variation within species is ubiquitous in both plants and animals. The native peoples of Alaska, Europe, India, and Africa differ in a wide range of characteristics, as do populations of the Bay checkerspot in California's outer Coast Ranges, inner Coast Ranges, Sierran foothills, and High Sierra. And all of these populations live in distinctly different habitats.

While there is a great deal of variation within species, the patterns of variation between species are usually characterized by wide gaps. To take an extreme example, both human beings and Bay checkerspots are geographically variable, but there is no overlap in their variation—no checkerspot would be mistaken for a human being and vice versa—and no one would question their distinctness. People and Bay checkerspots are different kinds of organisms by anyone's definition.

But things are not always so neat—an entire spectrum of degrees of difference between populations can be found. There are a great many examples of populations that have differentiated just by the frequencies of a few genes. Jasper Ridge and Sierran Edith's checkerspot populations are like that. Populations of other species have differentiated to a degree where they might or might not be considered separate species. The latter is the case with the European brown bear, the Alaskan brown bear, and the grizzly bear. Early taxonomists considered them each to be distinct species, but today taxonomists think that they all should be considered the same species. Still other populations are differentiated to the point where it is commonly agreed that they are separate species, but they still hybridize (mate and produce offspring) with one another quite easily. A well-known example of animals that have barely reached the status of separate species are the domestic dog and the coyote, which sometimes hybridize to form the "coydog." In the checkerspots of the western United States, virtually every degree of differentiation from "populations indistinguishable" to "distinct species" can be identified.

The point is that nature today presents a "snapshot" of an ongoing process that conforms precisely to the notion that differentiation of geographically isolated populations is the common mechanism of speciation. All over the world, in virtually all groups of organisms, populations are "caught" in that snapshot in every conceivable degree of differentiation.

This does not mean that all scientists concur on how this organic diversification takes place. Some evolutionists think that populations of organisms can diverge without geographic isolation. Also unsettled is whether eliminating migration between two isolated populations is more important in differentiation than their exposure to different selective pressures. This latter issue, like the neutrality controversy, is proving to be a tough nut to crack, since it is usually very difficult to measure either migration or selection in natural situations.

To ecologists, perhaps the most important controversy about speciation at the moment has to do with the *rate* at which it occurs in nature —or, more precisely, with the distribution of rates in time. Since Darwin's time, most evolutionists have thought that speciation goes on and has gone on pretty much constantly and gradually throughout the history of life. Recently, however, a group of paleontologists, led by Stephen Jay Gould of Harvard and Niles Eldredge of the American Museum of Natural History, has challenged this idea. Steve, who is widely admired for his brilliant essays interpreting science for laypeople in *Natural History* magazine, is also a formidable scientist. He and his colleagues interpret the fossil record as indicating that most speciation is concentrated into relatively short episodes in the history of life, and that between these episodes are long periods of evolutionary stasis in which species change but little and do not split into additional species. So, to Gould, Eldredge, and some others, diversifying evolution is more or less stalled in its tracks most of the time, proceeding in extremely short bursts between the long periods of inactivity. This implies that special ecological circumstances must arise periodically that encourage diversification—and since diversification seems to be going on at a rapid rate right now, it could mean that we are currently in an era that is ecologically unusual.

Furthermore, Gould and his colleagues more sharply distinguish the processes of microevolution, small-scale genetic changes within populations over short time spans such as the melanization of the peppered moths, from those of macroevolution, major changes through geological time such as the evolution of modern horses from their dog-size ancestors. Conventional evolutionists tend to view macroevolution as simply the cumulative result of microevolutionary processes continuing for long periods. Gould, paleontologist Steven Stanley of Johns Hopkins, and others invoke additional mechanisms that do not operate through changes in populations to explain macroevolutionary trends. One is selection operating between species (nonrandom, differential survival of entire species).

The view of Gould and his colleagues has become known as the theory of punctuated equilibria. One orthodox evolutionist, determined not to rise above *ad hominem* attacks, has labeled it "evolution by jerks." The more reasonable response of the conventional evolutionists, now called gradualists, to the punctuationists has been that conventional evolutionary theory had always recognized that there could be great differences in rates of speciation from group to group and time to time.

So which theory is right? Because evolutionary theory is so comprehensive and so well established, there is a predisposition among some evolutionists (myself included) to favor unorthodox ideas like punctuated equilibria. But, of course, being unorthodox is no guarantee of correctness in science. For every Galileo or Darwin who shatters the orthodox paradigm, there are dozens of Velikovskys or worse. Even when serious, talented, and respected scientists disagree with a commonly held view, they can be incorrect. For example, a determined challenge to orthodox speciation theory by some scientists who believe that in many cases speciation could occur without isolation has left the orthodoxy so far intact. Convincing evidence for the frequent occurrence of speciation without isolation has not been assembled. (This applies only to animals; speciation without isolation, involving chromosomal changes, has long been known to occur commonly in plants.)

The jury is not in on the punctuated-equilibria controversy. That the "snapshot" of differentiation we see today seems to reveal all stages of differentiation does not necessarily signal a win for the gradualists. We could, as I indicated, be in the midst of one of the periods of rapid speciation. And it is not fair to swallow the punctuationist view within the gradualist orthodoxy simply because the possibility of rapid speciation has always been part of that orthodoxy. The punctuationist view is about dominant patterns, not about what is possible—and it represents a genuine challenge to one widely held tenet within evolutionary theory.

There are, of course, numerous areas where evolutionary biologists disagree about the mechanisms of evolution. It is unfortunate, however, that these healthy disagreements about details of the process are all too often cited by creationists as indicating serious doubt among knowledgeable scientists about whether or not evolution actually occurred. Nothing could be further from the truth. My instincts and background tend to make me a gradualist (even though I root for the heterodox); Steve Gould is clearly a punctuationist. Yet we first met doing a radio show in which we both spent the vast majority of the program trying to persuade people calling in that there was no question whatsoever that evolution has gone on and that natural selection was a major factor in the process. Differences on evolutionary theory between Steve and me on one hand, and a "creation scientist" on the other, are roughly of the same order as differences on civil rights between a liberal democrat along with a middle-of-the-road democrat on one hand, and a believer in the divine right of kings on the other.

CHAPTER

3

SEX AND SOCIETIES

Behavioral Ecology

BENEATH THE sunbaked sandy soils of Kenya and the Horn of Africa live colonies of perhaps the strangest of all mammals—naked mole rats. They are virtually hairless creatures, with wrinkled pink skin and prominent incisors. At most five inches long, naked mole rats look curiously unformed, fetal or newborn—indeed, rather repulsive. But their appearance is not the strangest of their features.

The naked mole rats of a colony work in teams to excavate hundreds of yards of burrows. One mole rat works at the end of the burrow, chewing at the soil with its incisors. It then kicks the loosened earth to the animal just behind it. That animal backs along the burrow kicking the earth along with its hind feet and eventually deposits it in a side branch, where another mole rat is responsible for kicking it out onto the surface. The mole rat that has dropped its load meanwhile returns to the end of the new burrow. As it moves forward, it must straddle a chain of earth-laden mole rats backing away from the end,

81

Naked mole rats are burrowing mammals that have a social organization similar to honeybees. 1. An adult rat using its teeth in digging out a tunnel. 2. The "queen" in a colony just prior to the birth of twenty-three babies. 3. A weanling rat begging fecal material from an adult.

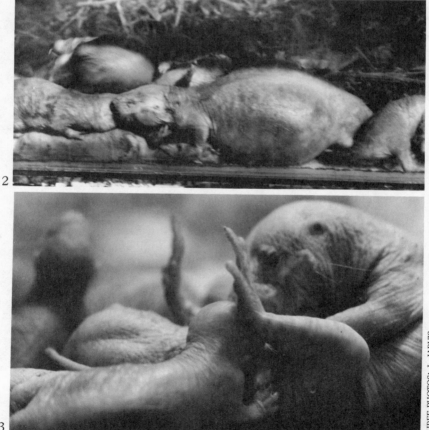

just as it had a few minutes before. In essence, the mole rats form a living bucket excavator, a stream of loaded mole rats backing up beneath a stream of unburdened mole rats moving forward.

But not all of the mole rats of a colony participate equally in these efforts. As Professor Jennifer Jarvis of the University of Cape Town has shown, mole rats are organized into a series of castes. One group, called frequent workers, does most of the nest construction and foraging for food. Another group, called infrequent workers, is composed of heavier individuals that do some work, but at less than half the performance level of the frequent workers. Mole-rat aristocracy is made up of even larger individuals that are nonworkers. They rarely do any digging or foraging and may be a reproductive caste, since male nonworkers are most likely to mate with the single breeding female (queen).

The queen produces up to four litters of as many as a dozen young annually. She also is the only female to suckle the young. All the other females appear to have quiescent ovaries and do not breed. Recent work by Jarvis and her student Brandon Broll indicates that the queen suppresses reproduction in the other females in part by producing a chemical messenger in her urine that is passed around the colony as the mole rats groom themselves after visiting communal toilet areas. Members of all castes, however, help to care for juveniles. For example, at three weeks the young are weaned, after which an important component of their diets (along with roots and tubers) is fecal material that they beg from adults.

Although the details are not yet fully understood, these amazing mammals appear to have evolved a social organization parallel to that of termites, bees, and other social insects. They have a division of labor based on castes, only a single reproductive individual in a colony, and several generations of offspring assisting the parent in the care of the young.

Mole rats also illustrate an important and obvious ecological principle. For sexually reproducing organisms—that is, for the vast majority of all organisms—one of the most important components of the environment is others of the same species but of the opposite sex. Nonsocial sexual creatures do not depend on members of their own sex, and depend on members of the opposite sex only for reproduction. Life expectancy, for example, of a single butterfly or rattlesnake is ordinarily as long as that of one in a group. In contrast, for social animals such as mole rats, other members of their own species of both sexes are the most important environmental factor. A lone mole rat, a lone honeybee, or (in most cases) a lone human being will not long survive.

Thus, just as sex and social behavior are pervasive concerns for most people, so they are central to the professional concerns of many ecologists. And for both ecologists and laypeople, interest in sex and societies usually extends beyond the bounds of *Homo sapiens.* The most basic social bond, that between mother and offspring, elicits empathy in most people, whether it is the humanlike interaction between a mother chimpanzee and her frightened offspring or the simple act of a female stinkbug guarding her young. The popularity of nature programs on television rests heavily on their portrayal of social interactions—male prairie chickens drumming to attract females to the mating ground, lionesses cooperating in the hunt, zebras or wildebeests taking advantage of the massed sensory apparatus of the herd to detect and thus avoid the lionesses. Who, in our overindustrialized and regimented society, can help but draw parallels (biologically naïve as they may be) between human life and the intricacies of the highly evolved insect societies? And who can avoid a small shudder of horror in the discovery that the ugly naked mole rats, mammals and therefore rather close relatives of ours, have traveled much the same ecological path as termites?

People want to understand their social nature and, above all, their sexual nature. Social scientists have long pondered such questions as what makes a good sexual relationship, or why do friends and lovers, parents and children, cling to one another. These are important questions. Ecologists often ask similar kinds of questions, but in different ways, ways that are intended to get us closer to first causes.

Let's start with a most basic question about sex. What good is it? Surely a silly question to anyone whose hormones are still circulating. But when they're not enjoying it, biologists wonder why it *is* so much fun, so much a focus of most people's lives. They puzzle over why sex, and its marvelous concomitant, sexual pleasure, have evolved.

As I mentioned in the previous chapter, in asexual reproduction, the offspring receive a set of genes identical to that of the parent, unless mutation intervenes. Looked at another way, the parent passes on *all* of its genes to each offspring. In contrast, sexual reproduction results in each parent passing only *half* of its genes to each offspring. Since the key to success in evolution is having one's genes represented in the next generation, asexuality would seem the ideal scheme. If a population had both sexual and asexual females, the genes of the latter ought to be passed on at twice the rate of those of the former—natural selection against the sexual reproducers should thus be intense.

Why, then, has sexual reproduction not disappeared from this Earth? Why, instead, is it by far the most common mode of reproduction? There is no one agreed-upon answer, although there is a consensus about where to look. You will recall that the feature that most clearly distinguishes sexual from asexual reproduction is that sexual reproduction results in recombination. The answer to the question "What good is sex?" therefore is generally sought in the advantages of recombination. Somehow, evolutionists reason, those advantages must outweigh the heavy disadvantage of passing on only half of one's genes at a time —what is often referred to as the "cost of meiosis."

But that's about where the consensus ends—the origin and significance of sexual reproduction remain two of the least settled topics in evolutionary ecology. There is no agreement, for example, on what kinds of environments favor the evolution of sex, or what ecological situations promote asexual reproduction. Controversy centers on the problem of how the heavy cost of meiosis can be more than compensated for by recombination. We know that the main result of recombination is the production of genetic variability in the offspring. We have already seen how important that variability is to the process of speciation. But how can it be selected for—that is, how can individuals that reproduce sexually pass on more of their genes than those that reproduce asexually, in spite of the latter's apparent mathematical advantage in the gene-reproducing game? Unless sex leads to more surviving offspring, sexual reproduction should not evolve.

One possible answer to this vexing problem is that producing an array of genetically variable offspring is a superior strategy in the face of environmental change or environmental diversity. A female lizard, for example, might produce ten young asexually. Each of her children would then have her full complement of genes, but all those genes might be lost if the environment changed so that individuals with her genotype could no longer survive. She and her offspring would all perish.

In contrast, if the female mated, each of her ten offspring would carry only a random selection of half of her genes combined with a random assortment of genes from the male, and they would be distributed in ten different genotypes. The same environmental change that made it impossible for her asexually produced offspring to survive would not necessarily be lethal to all of her sexually produced offspring. If even one of them could tolerate the new environment, then sexual reproduction would give her an evolutionary advantage. Similarly, if an environment varies from place to place, more individuals of a diverse group of ten young lizards may "fit in" than of a uniform batch of ten offspring.

This general answer to the question of why sex persists leans heavily on what is known as the "Red Queen Hypothesis." That hypothesis is named after the character in *Through the Looking-Glass* who has to keep running in order to stay in place. The Red Queen hypothesis basically holds that simply in order to persist, a population must continue to evolve—otherwise the population's ever-changing environment, especially its evolving predators and competitors, will force the population to extinction. Sex (and recombination) is seen as the only way that a population can keep running fast enough—especially to stay ahead of pathogens, which can rapidly evolve the ability to breach an organism's immunological defenses. Indeed, some ecologists think that the existence of disease may have been a crucial factor in the evolution of sex.

The variability-to-meet-change answer to the riddle of sex seems to receive some support from the water fleas (tiny crustaceans) and aphids. They commonly reproduce asexually in the spring and summer when conditions are relatively benign and predictable, and sexually toward the end of the growing season when, one might argue, catastrophic change is on the way. The precise time of winter's arrival, its severity, and the kind of spring that overwintering individuals will face are relatively unpredictable. But, then again, many other organisms that have multiple generations during a growing season—cabbage butterflies, for example—reproduce only sexually. And the whiptail lizards that reproduce asexually do not seem to live in environments any more predictable than those of their sexually reproducing relatives. Such are the difficulties of arguing about evolutionary ecology on the basis of examples. All too often, the counterexamples lie right at hand.

One additional problem with the notion of variability-to-meet-change is that if future environmental conditions are completely unpredictable, the variable-offspring strategy may not be best. Perhaps, for example, it would be best to produce identical offspring that are adapted to what, over time, have been the average conditions and which have a wide phenotypic tolerance to deviations from them.

A second possible answer to the question of why sex evolved is that, although it is disadvantageous for an individual, sex is advantageous for the population as a whole. A population containing many genetically different individuals is likely to persist in the face of environmental changes that would doom a genetically homogeneous one. This position invokes what is known as group selection to explain the predominance of sexual reproduction. The idea is that selection operating on populations has favored sex—those populations that by chance were founded by sexually reproducing individuals have been much more likely to persist than those populations founded by asexual individuals.

Group selection is often invoked to explain various kinds of altruistic behavior—behavior that promotes not one's own genes but those of others, such as one individual sacrificing its life or its opportunity to reproduce for that of another. Group selection has been considered by ecologists seeking answers about sexual reproduction because sex too is, in a sense, altruistic—a female halves the reproduction of her genes to collaborate with another individual in producing a population of offspring with desirable characteristics.

The problem with this scheme is that it does not explain how individual selection can be prevented from converting altruistic populations to selfish populations or sexual populations to asexual ones. Within any single population, individual selection favors those genotypes that are most reproductively successful. Therefore, in a population of altruistic or sexually reproducing individuals (favored as a population by group selection), individuals that, because of mutation or recombination, were selfish or reproduced asexually, would be at a great selective advantage. Thus, over time, the selfish or asexual individuals should completely replace the altruistic or sexual ones. There are very special circumstances in which group selection can work when it is opposed by individual selection, but most ecologists believe that it could not account for a phenomenon as widespread as sex.

The great British biologist-statistician Sir Ronald Fisher developed yet another explanation for sex. His idea was that sex (recombination) permitted favorable mutations to get together more rapidly than did asexual reproduction. With sexual reproduction, mutations occurring in two different females can, through recombination, end up on the same chromosome in one of their common grandchildren. But recombination can also rapidly move the mutant genes to separate chromosomes again; recombination is perpetually reshuffling the genes. It is generally accepted by theoreticians that Fisher's explanation is most likely to be correct for small populations in fluctuating environments—in which case it boils down to another expression of the variability-to-meet-change idea.

The truth is, no really satisfactory explanation for the evolution of sex and recombination exists as yet; both appear to be advantageous in some situations and disadvantageous in others. Since the overall question of why sex evolved remains a popular topic among evolutionists and ecologists, not much attention has yet been focused on some related puzzling questions, such as why a two-sex system is almost the standard. I say "almost" because in some protozoa and fungi there are recombination systems involving more than two sexes. In these species, several different types (sexes) of individuals exist that can mate ("conjugate") and exchange genetic material with those of other types but cannot mate with others of the same type. But why systems of multiple sexes have not become more widespread is unknown.

Given that there are normally two sexes, however, considerable progress has been made in understanding several other important issues, including the differences in the ratio of males to females in different species and the roles of sex in social organization.

Most people assume that equality in the numbers of the sexes, a 1:1 (male:female) sex ratio, is "normal," a natural result of the meiotic process that creates equal numbers of male- and female-determining genes. Equal sex ratios are obtained by one sex producing equal numbers of two types of gametes. A gamete is a mature germ cell (sperm or egg) carrying a haploid (single) set of chromosomes and capable of fusing with another gamete from the opposite sex to produce a zygote, a cell with a diploid (double) set of chromosomes which develops into a new adult. One type of gamete from the sex-determining parent produces a male zygote and the other a female zygote.

There is no reason for this equality in male- and female-determining gametes to persist, however. Genes that can distort the sex ratio are, in fact, rather common, and some species have many more females than males. Yet most species do have roughly equal numbers of males and females. Why?

Ronald Fisher had an answer for that one too. He considered what would happen if a population were genetically programmed to produce offspring in which males were twice as common as females. He reasoned that, in such a population, females would be twice as likely as males to find mates (if the population were monogamous, to take the simplest case, there would be mates for only half of the males). Thus any mutant individual that produced a high proportion of female offspring would be favored by selection because its offspring would be more likely to reproduce than one that had a high proportion of male offspring. Its genes would thus be more likely to be passed on to the generation of its grandchildren.

Thus parents that produce more offspring of the rarer sex in a population generally will have a selective advantage over those that produce more of the commoner sex. Their children, with fewer individuals of the same sex to compete with them for mates, will leave more offspring of their own. This will increase the frequency of individuals that are genetically programmed to have more offspring of the rarer sex —until that sex is no longer the rarer one. This tends to stabilize the sex ratio around 1:1, for whenever one sex begins to fall behind (becomes rarer), individuals that favor that sex among their offspring would leave more descendants with a resulting increase in the frequency of the rarer sex. Thus, in this situation, whether or not parents with a hereditary disposition to produce children of a given sex are favored by selection depends on the frequency of that sex already in the population.

But Fisher's actual model is more complex still. He argued not that the sex ratio itself would be even but that parents would expend equal amounts of energy producing males and females. To understand what this means, let's look at the reproduction of the European sparrowhawk. Females of this bird are twice as heavy as males when they leave the nest. It is probably safe to assume that it takes roughly twice the parental effort to rear a female to the size at which it can fledge as to rear a male. Now suppose the sex ratio in a sparrowhawk population were 1:1, and offspring of both sexes had equal opportunities to mate and produce offspring themselves. In that case, selection would favor genotypes that produced males, the "cheaper" sex. A parent that fledged four males would (on the average) leave twice as many grandchildren as one that fledged two females, even though it took the same amount of effort to produce the pair of females and the quartet of males. A female-producer would have to invest twice as much effort to have the same reproductive success as a male-producer. Equal investment would not result in equal chances of having grandchildren.

Selection favoring male-producers would gradually increase the frequency of male sparrowhawks until the sex ratio was shifted to 2:1 in favor of males. At that point, a parent that fledged four males would have exactly the same chance of producing grandchildren as one that fledged two females. That is because the female offspring, now outnumbered two to one by males, would be twice as likely to find a mate. Equal investment (four males or two females) would now result in equal chances of having grandchildren.

Makes sense, doesn't it? Remember, however, that the standard for a theory in science is that it not only must make sense, it must also stand up to experimental test. Counts of young sparrowhawks in England by I. Newton and M. Marquiss of the Institute of Terrestrial Ecology in Edinburgh revealed 1,102 males and 1,061 females—hardly the predicted 2:1 ratio. Back to the drawing board. It turns out in this case that our assumption was wrong. The sparrowhawks actually invest about equal amounts of food in the two sexes before they leave the nest—the females get heavier but the males use the energy to develop faster and fledge first. So a roughly 1:1 ratio fits the theory (once we get our assumptions right).

A study of grackles also seems to confirm the theory. In grackles, a kind of blackbird, females are smaller at fledging, and the sex ratio is, as predicted, biased toward them. Unfortunately, in this case, data on actual investment are lacking, so the final word is not in.

Investment theory is actually more complex than I have indicated. For instance, in certain circumstances involving inbreeding and great individual-to-individual variation in reproductive success, biased sex ratios are both predicted by theory and found in practice. But many more tests, using a wide range of organisms, will be required before ecologists will be satisfied that nature and theory are in harmony.

There are other mysteries about sexual reproduction. One is why the evolutionarily correct ratio is achieved in one way in one organism and another in a different organism. Human males, for example, determine the sex of their offspring by producing sperm carrying either an X or a Y chromosome. An egg fertilized by sperm with an X chromosome becomes a girl, and one fertilized by a sperm with a Y chromosome develops into a boy. But the system is not the same in all organisms; in checkerspot butterflies, it is the chromosome complement of the egg, rather than the sperm, that determines the offspring's sex.

Why hasn't evolution produced a system in which, say, there are two kinds of males—one that produces only male-determining sperm and the other female-determining sperm? Why don't more organisms change their sex in response to environmental conditions, as do some

SEX AND SOCIETIES 91

coral reef fishes that have only one male per family group? If that male is killed and another male doesn't take over his harem, the largest female changes into a male. Small wonder that so many books have been devoted to the biology of sex—and many more will be.

The key point is that sexual reproduction, like all other features of living systems, is not a "given." Rather it is a result of billions of generations of organisms being shaped by their environments in an evolutionary process. It will take quite some time to answer questions as diverse as why sex occurs in life-forms as different as bacteria and people, and whether or not there are ecological and evolutionary reasons for different patterns of sexual behavior in men and women. But enough has already been learned to assure us that an ecological approach to sex and society can yield valuable insights.

Now let's turn from the puzzles of sex and consider the arena within which most sexual activity among animals occurs—the social group. The distinguished Harvard evolutionist E. O. Wilson has defined a society simply as a group of animals of the same species organized in a cooperative manner. This simple definition encompasses many kinds of societies, and the variety of social behaviors found within them is enormous. Elephants attempting to help an injured comrade and male elephant seals guarding harems are behaving socially. So are herrings zipping about in a school. Social behavior includes dominance hierarchies ("pecking orders") in coral reef fishes, chickens, and human beings. Lions, wolves, and killer whales use social cooperation in hunting; zebras, crows, herrings, and water striders employ social behavior in defense against predators. Frogs' croaking, birds' singing, termites' swarming, and people's courting are all social activities.

How is it possible to make sense of these behaviors, as well as the vast number of other ways that individuals within a population relate to one another? Faced with seemingly strange social behavior, one possible way to understand it is to determine how it evolved—to find out why the environment favored the reproduction of individuals that performed the behavior over those that did not.

Our exploration of social behaviors will start with a common and seemingly simple one, the formation of groups—schools, coveys, flocks, herds—by nonpredatory animals. Why would such behavior evolve? Naturalists have long assumed that, especially in open country, groups evolved simply because there is safety in numbers. When danger threatens, animals tend to bunch up. If a hawk approaches a flock of starlings,

A social hunting group—members of "J pod" of Puget Sound orcas (killer whales). The individuals shown are all females or young; adult males can be distinguished by their much taller dorsal fins, and may weigh as much as eight tons. The pod, which is thought to be an extended family, moves and hunts together. In the Antarctic, Anne and I watched a small pod repeatedly smash into a pan of ice on which a Weddell seal was sheltering, washing water over it, until the seal was dumped into the sea and devoured.

they cluster more closely together. A threatened school of fishes or herd of sheep, deer, or musk-oxen closes rank.

But why should there be safety in numbers? Why shouldn't a tightly packed flock simply be a conveniently laid table for a pack of carnivores, saving it the trouble of assembling its food from scattered individuals? W. D. Hamilton, one of the most insightful of modern ecologists, developed a simple model of why prey in the open should cluster together. He called it the "geometry of the selfish herd."

Hamilton's basic assumption was that a predator would tend to attack the nearest available victim. It would then behoove any potential prey individual to approach another, because two standing side by side would each have half the "domain of danger" of a single individual (since the other would be closer to a predator approaching from half of the possible directions). A group of four would again halve the domain of danger of each, and so on.

In this model there is no altruism. Each individual is attempting to save its own skin by placing others between itself and the predator. In fact, attempts of individuals to bury themselves in the group have often been observed in potential prey. Outside fishes attempt to penetrate into the school; sheep harried by a sheepdog try to butt their way into the flock. This behavior makes sense, as there are numerous observations indicating that predators are much more likely to take peripheral or laggard individuals from herds and schools. In the selfish herd, the evolutionary advantage goes to those fast, strong, or wise enough to get to the inside.

Although it may give some individuals advantages over others, flocking behavior, even in the open, may work to the disadvantage of the group as a whole. It may make prey easier to locate, the flock being more obvious than solitary individuals. The confusion of a milling herd may make the predator's work easier, not more difficult. Indeed, some predators have become specialists in attacking groups. For instance, swordfishes and sawfishes appear to be specialized for feeding from schools. They lunge for the center of the group and kill large numbers of individuals by slashing through the school with their "weapons."

As a result, natural selection may favor aggregation in some circumstances and not others. There may be an urge to associate at some seasons of the year and not at others, depending, for instance, on the amount of cover available, the activities of predators, and the stage of the reproductive cycle. For example, males of waterbuck and some other African antelopes live solitary existences during the mating season but form bachelor herds at other times of year.

There is other evidence, however, that flocking can work to the benefit of the flock as a whole—that is, there are factors that protect *all* individuals in a group of prey. Some of that evidence consists of observations of attacks on groups, especially of predators on schooling fishes, and on the experience of human hunters trying to shoot birds or catch butterflies. These indicate that a large group is such a confusing target that it is quite difficult to select and bag a single individual. Many a hunter has blasted into the middle of a flock of ducks only to discover that he has missed them all. Whether the confusion of the herd benefits hunter (as mentioned above) or hunted doubtless varies with the circumstances.

Another advantage of being in a group is often illustrated by fishes. A tightly packed school of a harmless species may simulate a single large predator—as I learned to my fright one day while scuba diving in

Bora Bora. It only took a moment for me to realize that a mass moving toward me was a school of harmless mullet rather than a large shark, but had I the reaction time of a middle-sized mullet-eating predator such as a grouper, I would have been long gone before resolving the true nature of the apparition.

Another defensive advantage of a group is that the senses of numerous individuals are available to detect the approach of predators. In a simple model, suppose that a certain species of bird must look for food half of the time in order to stay alive. Suppose also that it can spend all the rest of its time watching out for predators. A cat might easily sneak up on a single bird of that species, moving only during the 50 percent of the time the bird is feeding. A flock of ten birds is an entirely different story, though. If each bird spends half of its time watching, even if there is no relationship between when one bird watches and another bird watches, there will be at least one bird watching more than 99.9 percent of the time.

A little arithmetic shows why. If there are two birds, each will *not* be watching half of the time. Since the behavior of each (by my definition) is independent of the others, the proportion of the time when both will not be watching is $\frac{1}{2} \times \frac{1}{2} = (\frac{1}{2})^2 = \frac{1}{4}$. Similarly, the time when none of ten birds will be watching is $(\frac{1}{2})^{10} \times 1/1024 =$ less than .001. Thus 99.9 percent of the time at least one bird is watching.

As these calculations indicate, however, assemblage into larger and larger groups provides smaller and smaller increases in security as long as each individual spends a substantial portion of its time on the lookout. In the case just given, doubling the flock size to twenty increases the amount of time at least one individual is watching by less than a tenth of a percent. Of course, there may be other advantages to such an increase in flock size: larger assemblages could allow individuals to spend less time on sentry duty and more time feeding.

Various experiments do, indeed, support the notion that increased flock size improves predator detection. For instance, a hawk model was flown over different-sized flocks of wild doves, and the results were recorded with a movie camera so that the time taken to alert the flock could be precisely measured. Between flock sizes of four and fifteen, there was a clear relationship between the number of birds and the speed with which they were alerted. Furthermore, the evidence indicates that in larger flocks, as one might expect, individual birds actually do spend more time feeding and less time on lookout.

Of course, in order for one individual in a flock to alert others, it must communicate to them that there is danger. Social animals have evolved many ways of communicating with one another—not just about danger but about the presence of resources, the availability of mates, and many other things. For most, sight and sound are the preferred media, but other, less familiar forms of communication are also used. Perhaps the most unusual are the dances used by honeybees to communicate the distance, direction, and quality of a food resource. The dance is done in the darkness of the hive, and the information is transferred by touch and smell.

When it comes to warning others of danger, however, sound seems to have many advantages. The recipient does not have to be facing the sender of the message in order to receive it. And a brief warning call does not attract as much attention to the sender as would jumping up and down or running in circles. In fact, many warning calls have qualities that make it difficult to locate the source—and may even be ventriloquial, giving a false impression of the position of the animal giving the call, thus reducing the chances of its being found and eaten because it gave its presence away by sounding the alarm.

Still it is difficult to explain why an individual should take *any* risk in order to alert other members of the group. Even incurring the smallest additional chance of attracting an approaching predator's attention would appear to fly in the face of individual natural selection. Why not just sneak quietly away, leaving your buddies to feed the predator, and donate *your* genes to the next generation? Could it be that group selection is the answer—that flocks containing altruists persist and those that do not, perish?

But the difficulty with group selection remains. Favorable as certain behaviors are for the group, they will be selected against at the individual level if those that perform them are more likely to perish than those that do not. In a flock consisting of all altruistic callers, any mutant or recombinant individual that keeps quiet and sneaks away will be at a selective advantage, and its kind of genes will soon be the only ones around.

The role of group selection is not the only issue in the debate over alarm calls. There is also considerable controversy about whether the caller actually is being altruistic at all or perhaps is just manipulating the group in a way that enhances its own chances of survival—for example, causing the flock to spring up between itself and the advancing predator. Behaviorist Kenneth Armitage, on the basis of his work with marmots, takes the middle ground. He believes that many alarm calls entail no risk to the callers. They simply put the predator on notice that it has been spotted, allowing it to waste no further time stalking and allowing the marmot to return to the business of eating.

But, assuming that at least some alarm calls are altruistic, is there any way that such behavior could have evolved? We again owe the best evolutionary explanation so far to W. D. Hamilton. He put forth in 1964 a notion that has had profound importance for the understanding of much of social behavior—kin selection. Like many important ideas, that of kin selection is disarmingly simple in retrospect: an individual may promote the reproduction of its own kind of genes by helping its relatives (kin). When the relatives are the individual's own children, this is obviously just basic natural selection. But it also can work when other relatives are involved.

To see how kin selection can result in altruism, one must differentiate between altruism toward just anyone and altruism toward one's relatives. An individual of a sexually reproducing species shares half of its genes with its brothers and sisters. You can see this intuitively by imagining that if Sister received a copy of gene A^1 from Mom, who has the genotype A^1A^2, then Brother has a 50:50 chance of getting a copy of A^1 also, since both Sister and Brother must get either A^1 or A^2 from Mom.

One half also turns out to be precisely the proportion of genes shared by an offspring with each parent in a sexually reproducing species; the other half it shares with the other parent. (In asexual species, of course, parents and offspring share, ignoring mutation, 100 percent of their genes.) It can also be shown that grandparents and grandchildren share 25 percent of their genes; first cousins, 12.5 percent; and so on.

Suppose, then, that an individual could sacrifice its own life in order to save the lives of three of its siblings. Or take, say, a 5 percent chance of being eaten to save a first cousin. Such altruistic behavior should be favored by selection, since in each case more genes identical with those belonging to the altruist are saved for posterity than are lost.* Furthermore, individuals that have ceased reproduction should, if genes ruled all, be willing to make virtually any sacrifice that promoted the survival or well-being of their grandchildren or other reproductive rela-

---

* This example is oversimplified since, to be precise, one must consider the reproductive potential of the altruist. If the odds were that the potential altruist could produce twenty more offspring before dying if it did not sacrifice itself, then the altruistic behavior would be selected against.

tives (by definition, kin selection is not involved when direct descendants are aided).

Hamilton's ideas have thus broadened our previous concept of fitness, which you will recall is the proportional contribution of an individual's genes to posterity, to encompass what is called inclusive fitness —the contribution of an individual's genes and those genes of its kin that are identical. Behavioral ecologist Paul Sherman of Cornell University, testing Hamilton's idea with observations of Belding's ground squirrels, found evidence that alarm calling is an example of altruistic behavior produced by kin selection. Female squirrels gave many more alarm calls on spotting predators than did males, and females with young more often gave the calls than those without. Female Belding's ground squirrels are much more often associated with kin than are males, and females with offspring have the most close relatives to protect. Thus the simplest explanation for Sherman's data is that while alarm calling did increase the risk to the caller, for those able to alert relatives, the individual risk was compensated by increased inclusive fitness.*

So much for social behavior in prey animals. Do predators sometimes benefit from being in groups? Many animals—from tigers, koalas, marmots, snakes, and most hawks and fishes to butterflies, dragonflies, and ladybird beetles—seek their food in solitude. Some even defend their feeding grounds against other individuals of their own species. Others, including many songbirds, Eleanora's falcon, surgeonfishes, hyenas, lions, and people, often forage in groups. In most of the latter species, one does not have to look far to discover an individual selective advantage to account for the evolution of the behavior: predators hunting in groups often get more to eat than loners do.

Hyenas, for example, often hunt in groups. (Hyenas, by the way, have a well-entrenched but undeserved reputation as cowardly scavengers. They are actually bold hunters, kill most of their own food, and will even chase cheetahs off of cheetah kills.) Hans Kruuk of Oxford

---

* Although kin selection and inclusive fitness are excellent heuristic devices, it has been shown that any argument that can be made in terms of inclusive fitness can also be made in terms of plain old Darwinian fitness. To do so, the mathematical treatment must be constructed so that the Darwinian fitnesses of genotypes are themselves functions of the other genotypes in the population. In other words, one individual's fitness depends on the kinds of individuals (including proportion of relatives) with whom it lives. An alarm caller will be more fit in a population consisting mostly of close relatives than in one consisting mostly of unrelated individuals.

University showed the advantage of group hunting in his classic study of hyenas. Each of two hyenas got more to eat by hunting cooperatively than by hunting alone. In teams they were much more successful than single hunters at capturing wildebeest calves. The mother wildebeest could protect her calf effectively against a solitary hyena, but not against the teamwork of a pair. Similarly, George Schaller and others have shown that, for lions, the pair seems to be the ideal-sized hunting group.

Ground-feeding birds in flocks benefit by flushing prey for each other; the startled insect that escapes the beak of the startler is often snapped up by another member of the flock. Other birds may cooperate in the air. Ecologist Hartmut Walter's detailed studies of Eleanora's falcons around their Mediterranean island breeding grounds showed that they form "falcon walls"—living nets of hovering hawks that wait for migrating birds traveling from Europe to Africa. A small bird fleeing from one falcon will often be taken by the next one along the "wall."

A standard hunting technique of our own ancestors was for groups to drive herds of large animals over cliffs or into water where they would die from the fall or drown, or be killed by the hunters. Eskimos, for example, would build funnel-shaped arrays of man-size stone cairns leading to a cliff or other natural trap and then herd caribou into the mouth of the funnel. Women and children behind some of the cairns would wave their arms and shout, giving reality to the rows of surrogate people and hastening the caribou down the funnel to their doom. This technique, in essence, permitted small groups of Eskimos to enlarge their numbers (from the caribous' perspective) and increase the success of their hunting.

Herbivores as well as carnivores can gain an advantage from foraging in groups that is unrelated to their greater safety from predators. For example, "herds" of grazing surgeonfishes or tang, whose striped livery has earned them the common name "convict tang," are a common sight on Pacific coral reefs. The tang swim three or four feet above the reef and periodically swarm as a group down to the reef to gobble up algae. Their feeding bouts are vigorously opposed by damselfishes defending territories on the reef surface; but, at any given point, the tang so outnumber the defenders that the latter's efforts are for naught—the tang easily feed as a school in circumstances where individuals, harassed by the aggressive damselfishes, would find grazing all but impossible.

But why are the damselfishes defending portions of the reef? Territoriality, the defense of all or part of the area where an individual goes about its normal business, is one of the most widespread and intensively studied ecological phenomena. Everyone who has ever heard a bird singing has at least had contact with it.

For a long time, territoriality was regarded as defense of an area only against members of the same species (intraspecific territoriality). Some biologists thought it had evolved as a population-regulating mechanism—since the availability of space for territories puts a lid on population size. Although it often seems to fulfill that role, it is unlikely to have evolved for that reason. If behaviors evolve that benefit the population and not the individual, then group selection must be involved. But, again, group selection is unlikely to be the reason for the evolution of a phenomenon as common as territoriality.

The current view is that territoriality evolved to permit individuals to protect or monopolize resources (especially food) or mates, or a combination of these. The damselfishes are defending the areas in which they feed, for instance. With respect to mates, territorial defense clearly only has to be against members of the same species. A male meadowlark's opportunity to find and keep a mate is not threatened by the proximity of a male robin. Similarly, the closest competitors for food sources are also most likely to be members of the same species. Hyena clans studied by Kruuk in the Ngorongoro Crater of Tanzania defend large territories against other clans. All the clans not only hunt the same prey, but members of one clan will attempt to sneak in and feed at another clan's kill.

Studies of the variation in intraspecific territorial behavior under different conditions support the belief that resources often play a major role in determining that behavior. Research by ecologist Frank Pitelka of the University of California at Berkeley and his colleagues has shown, for example, that the territory sizes of arctic sandpipers were smaller in parts of the birds' breeding range where the growing season was longer and the food supply was more abundant and dependable than in areas with a shorter season and less food. Apparently, in the plusher regions of the sandpipers' distributions considerably smaller territories nonetheless contained sufficient resources for successful breeding.

Interspecific territoriality (an individual of one species defending a territory against individuals of other species) often appears when different species are using the same resource. The damselfishes that live on the surface of the reef feed in part on the algae that are grazed by the convict tang and in part on a variety of small invertebrates. They defend their territory against virtually everything that moves. Even though they are only a few inches long, they will attack the toes of a six-foot-tall human diver with the same ferocity as they would another damselfish. Apparently, any animal not clearly a dangerous predator is considered a threat to the resources of these small omnivores.

When birds defend territories against other species, they show more discrimination than the damselfishes and normally defend only against species very similar to themselves. Since the birds' territories are comparatively large, and the potential resources in them more varied, the territory holder can save a great deal of energy by attacking only those intruders that most directly threaten the resources it uses.

Whatever the reason for the territory, animals often have evolved "keep out" warnings that alert others to its boundaries. These warnings presumably reduce the defensive effort of the holder by reducing the number of invaders; many potential intruders heed the warning and save energy themselves by staying away and avoiding a confrontation. Underwater, these signals are usually visual and may simply consist of the territory holder making his presence obvious. Alert damselfishes hanging in the water above the reef are advertising their willingness to defend plots in a bed of staghorn coral. Or signals may be more elaborate. I have been treated to the territorial display of an eight-foot-long reef shark, a little "dance" fifteen feet or so in front of me with back arched and pectoral fins extended. Able to take a hint, and not anxious to participate in an episode of natural selection, I backed away—saving us both the energy expenditure of an actual encounter.

On land, where odors can linger, many mammals, from mice and rabbits to wolves and hyenas, use scents from urine, feces, or special glands to mark the boundaries of their territories. Recently, Bert Hölldobler and E. O. Wilson have found that certain species of ants use scent to mark territories and to rally nest mates to defend the nest when it is threatened by intruders.

The most familiar territorial signals, of course, are the songs of male birds. Although the songs were once thought to have been created for the entertainment of *Homo sapiens* and to express inherent joy, ornithologists have long been convinced that at least one of their functions is to warn off other birds. Experimental verification of that hypothesis is surprisingly recent. J. R. Krebs of Oxford showed that when territorial male great tits (European relatives of American chickadees and titmice) were removed from their territories, the newly vacated areas were quickly occupied by other males that had not previously held territories. But, when the empty territory was "defended" by a loud-speaker singing songs of male great tits, it was occupied much more slowly than if it were empty or defended by a loudspeaker playing a note on a tin whistle (the "control" to make sure that neither loudspeakers or just plain noise would discourage other tits from occupying the territory).

It has also been shown that the more great tits there are in an area, the smaller the reproductive success of the average bird. One would therefore expect individual males holding territories to attempt to mini-mize the number of other territory holders in the vicinity. And con-versely, given a reasonable choice, territory seekers should shun areas already heavily occupied. That seems to explain why individual great tits sing several different kinds of songs and often change their tune when they fly from one perch to another—they are trying to give the impression that they are more than one individual.

Krebs came to this conclusion when he found that the most effec-tive loudspeakers in his experiments were those that played not a single song but a repertoire of several great-tit songs. Krebs called the notion that territory holders were trying to fool potential territory holders the "Beau Geste Hypothesis." Movie buffs will remember the dead legion-naires of Fort Zinderneuf propped in the embrasures while Gary Cooper ran from corpse to corpse firing rifles in an attempt to fool the attacking tribesmen into thinking the fort was fully manned. The ruse worked well for Gary, but finding out whether the hypothesis explains Krebs's results will require further experiments.

Territorial behavior can be fascinating to observe. My introduction to this sort of behavior came as part of a research project that was fated never to be completed. A couple of decades ago, two Stanford col-leagues, Ben Dane and Richard Holm, and I decided to study the effect of a starling invasion on the number and size of meadowlark territories on Jasper Ridge. The starlings were just moving into the area, and it seemed like an ideal opportunity to take advantage of a natural experi-ment.

We mapped the meadowlark territories with the help of a moth-eaten stuffed male meadowlark from David Starr Jordan's half-century-old small-bird collection. (Jordan, the father of American ichthyology, the study of fishes, was Stanford's first president. Among Stanford biologists, he is perhaps best known for his statement: "Every time I remember the name of a student, I forget the name of a fish.") The procedure we used was simple. The stuffed male, known as George, was set up in various places in the Jasper Ridge grassland. Each time, it was immediately attacked by a male meadowlark, which would rush at it from a hundred feet away or more. Even in his dilapidated condition, long-dead George was perceived as a rival. Indeed, I suspect a two-dimensional model painted with a black V on a yellow chest—the characteristic meadowlark marking—would have elicited the reaction. Each time we set George up, we noted which particular meadowlark was attacking him and then quickly intervened and moved an increasingly threadbare George to another position. By repeatedly moving him and observing his attackers, we were able to define the boundaries of each meadowlark territory.

But the starlings moved into the area much more slowly than we expected, and our initial enthusiasm waned. Shortly thereafter I went on sabbatical to Australia, leaving Dick with part of my teaching burden, Ben left Stanford, George went back into the collection, and the meadowlarks were left in peace. The response of the meadowlarks, however, made a deep impression on me. I'm reminded of George whenever I hear a meadowlark's distinctive song, and I've had an interest in territorial signals ever since.

Serendipity later got me involved in another study of territoriality, this one concerning the nature of territorial signals. It all began when, in 1965, my wife, Anne, and I decided to take a ship to Australia, where I would be spending my sabbatical; we wanted to feel we had really traveled a long way, rather than just having been bounced around for a few hours in an aluminum tube. During our trip, I was introduced to snorkeling on the coral reefs of Bora-Bora by Hal Wagner, the chief purser on the Matson liner *Mariposa,* and I was instantly hooked on the spectacularly colored reef fishes. That hook was later set at Heron Island on the Great Barrier Reef by Frank Talbot, an ichthyologist at the Australian Museum, whom we had met previously when he was a postdoctoral fellow working on Jordan's fish collection at Stanford.

The beauty and ever-changing action of the reefs were too great a lure. Anne and I a few years later became SCUBA certified and began to do research on reef fishes whenever the opportunity arose. We had also fallen in love with Australia, and in 1974 we made plans to return. One of our main goals was to visit the Australian Museum's Lizard

Island Research Station on the Barrier reefs. Frank was by then the director of the Australian Museum. Frank and I had wanted to do research together for a long time, and the planned visit to Lizard Island seemed like the ideal opportunity. But what to do in a single short period in the field? A study related to territorial signals seemed ideal, since it did not necessarily require observations done over a long period.

One of the dogmas of reef-fish ecology, tracing to the great Austrian ethologist (student of behavior in natural environments) Konrad Lorenz, was that the spectacular colors of reef fishes were territorial warnings. Lorenz called the markings "poster colors" *(Plakatfarben)* and considered that all poster-colored fishes were territorial and all territorial fishes were poster-colored. This did not jibe, however, with my casual observations. The most vigorously territorial fishes that I had encountered were the dull-colored damselfishes. And some of the most spectacularly colored ones were the butterfly fishes of the Pacific, which I had repeatedly observed swimming in schools—not the sort of behavior one would expect of territorial animals.

So Frank and I decided we would put Lorenz's poster-color hypothesis to a simple but nonetheless systematic test. We would use several species of Australian butterfly fishes that showed dramatic examples of poster coloration for our experiments. Frank had some fine life-size models of the butterfly fishes made, painted in a lifelike pattern on one side and white on the other. These were then attached to long Lucite wands so that either side could be presented to a fish by one of us holding the other end of the wand. With the models and two of Frank's young colleagues, we flew to Lizard Island.

It was a great field trip, involving diving much of the day and often at dawn and dusk as well.* The most interesting thing we found was that the presentation of models to the butterfly fishes did not produce any reactions remotely resembling the responses of the meadowlarks to George. The fishes usually approached but did not attack the realistic side of the model, and occasionally attempted to school with models of the same species. And they showed little or no reaction to the white (control) side of the model.

Individuals of only one of eight tested butterfly fish species as-

---

* Dawn and dusk diving is necessary because much of the action on the reef is during the change-over from the day to night shifts—quite different fishes are active at the two times—and territorial behavior might occur during the change-over that would not be seen at other times.

saulted the models, attacking most vigorously when the model (of a different species) was brought close to the flat-topped plate coral with which the attacking species is closely associated. In no case did the appearance of a model bring an individual dashing to the attack from any distance, as the meadowlarks did when confronted with George. Systematic observations of interactions between the butterfly fishes also produced little evidence that their poster colors served as territorial signals. During the day, in fact, they usually were distinctly nonterritorial, roaming widely over the reefs, usually singly or in pairs, although groups of up to thirteen individuals were observed.

The main territorial interactions that we observed were at dusk as the fishes settled down into their night resting places. We could identify many individuals by small variations in their markings, and we soon learned that those fishes that we could recognize (and most of the others, we assume) slept in the same places each night. Some butterfly fish species were very aggressive in defending those resting places, chasing away other individuals. Other species roosted communally in coral heads, with some jostling and nipping of smaller individuals by larger ones as they settled down. But most of these interactions occurred at close range—a "territory" in a roost was not more than a couple of inches across. And all of these interactions occurred *after* the bright poster colors had faded to duller nocturnal patterns.

Thus the Lorenz hypothesis was pretty well falsified. In the most vividly poster-colored reef fishes, the colors clearly do not serve primarily as territorial signals. But if they were *not* territorial signals, why had the poster colors of reef fishes evolved? Part of the answer to that question seems to be for species recognition; that is, poster colors may be important in helping members of pairs or of larger groups to keep sight of each other. The prominent colors may also be a reminder to predators that have tried before to catch them that butterfly fishes are hard to catch or too spiny to swallow easily. Or they may serve as a different sort of predator defense. The fishes are very conspicuous when they present their poster-colored broadsides. But when they turn to flee, they present their very narrow back/tail view to the predator and seem to disappear. Predators may find this sudden "disappearance" sufficiently confusing at the moment of attack to give the butterfly fish a chance to escape. Thus, what Lorenz thought had a single function probably has several.

While there may be aggressive territorial interactions between individuals and between groups, a first impression would be that when nonterritorial individuals occur together there is relative peace within the group. With the exception of occasional squabbles, usually of short duration, a pack of wild dogs, a flock of chickens, or a grazing school of tang would seem to be a most unaggressive assemblage. In the 1920s though, a Norwegian scientist, Thorleif Schjelderup-Ebbe, carried out some pioneering experiments with flocks of hens that put to rest this idea of easy interindividual relationships among vertebrates. He and his successors have been able to show that peace reigns in *established* flocks because the individuals have already fought it out and determined who can best whom in the scramble for chicken feed. Once the "peck order" has been established, the chickens remember it—each hen knows whom she can dominate and who will dominate her. If a new hen is added to the flock, she participates in a series of battles until her relations with all of the other hens are established.

In flocks of hens, it is the peck order (a form of what is more generally called a dominance hierarchy) that maintains social stability. Those near the top in the hierarchy almost certainly have increased fitness because of their superior access to food, nesting sites, and so on. The more submissive birds, while their access to food is not the best, are at least able to enjoy the benefits of group living without chronic harassment by bullies.

Dominance hierarchies are a widespread social phenomenon among the vertebrates, although they vary greatly in their strength and organization. For aggressive mammals such as the hyena, the existence of a well-established dominance hierarchy seems to be the only way the animals can associate with one another without mutually destructive results. Appeasement gestures using the penis are extremely important in the maintenance of peace in hyena societies. Indeed, they are so important that female hyenas have evolved a pseudopenis, which is almost indistinguishable from the real thing, so that they too can make these socially indispensable gestures. Some say this is the source of the ancient notion that the "laugh" of the hyena is a sign of its pleasure over being able to "change sex."

In some groups, the existence of dominance hierarchies in nature is only suspected. For example, subsequent to our original work on the Great Barrier reefs, Anne and I have shown mirrors to coral-reef fishes in their natural habitats. This sometimes elicited violent attacks on their mirror images from otherwise seemingly peaceful individuals of nonterritorial species. Our results suggested that dominance hierarchies are more common in reef fishes than previously suspected. Often attacks on their mirror images were made simultaneously by two members of the same species, which ignored each other. Presumably, each knew the other's place in the hierarchy, but was attempting to establish its rank relative to the "stranger" in the mirror. Each fish interpreted its own image as the equivalent of a new hen in the flock. If they had had the mental capacity, the combatants would surely have been astonished at the persistence of their adversaries. Each one certainly seemed puzzled when its opponent disappeared every time the fish carried the battle beyond the borders of the mirror. Often only the removal of the mirror from the water after an hour or so would end the "combat."

These studies provided an interesting bonus. Generally, the fishes made a persistent head-on biting assault on the mirror. One species of surgeonfish, however, made repeated swipes at the mirror with the bony "scalpels" that are found on either side of the base of the tail (and give the group their common name). This confirmed a long-held suspicion of ichthyologists that the scalpels played a role in intraspecific interactions as well as possibly being defenses against predators.

Territoriality and dominance hierarchies are both interwoven into the social life of some reef fishes, as illustrated by the cleaner wrasses, small fishes that make a living by gleaning parasites and dead scales off other fishes. In the most widespread cleaner, called the "paradise fish" in Tahiti, the males are highly territorial and are accompanied by a harem of females structured by a dominance hierarchy. In a famous study, D. R. Robertson showed that when the male died, there were two possible outcomes. One was that a male holding an adjacent territory would move the widows into his own territory and harem. The other was that the dominant female would quickly change into a male and take over the harem before another male could. In fact, all males of the paradise fish were previously dominant females. In these fishes, who gets to breed, and even whether an individual breeds as a male or female, is determined by a combination of territorial and dominance behavior.

How mates are obtained, the formation of pairs or other reproductive groupings, and the patterns of parental care of offspring are known collectively as an organism's mating system. Animals display an astonishing variety of mating systems—as the cleaner wrasses' system suggests. Evolution, of course, has shaped those mating systems, just as it has shaped the sexual mechanism itself.

A key to understanding mating systems is the greater investment of energy that females always put into each fertilized egg and usually put into raising each offspring, compared to males. The greater investment at the egg stage is inevitable because females produce larger gametes, and thus must expend more energy to produce each one. A human egg, for example, is thousands of times bigger than a human sperm. You might wonder if females *always* produce larger gametes. I can reply "yes" with an assurance rare for a response to a biological question, because that is how male and female are defined—by the relative sizes of their gametes. The asymmetrical investment of the two sexes often becomes still more uneven when parental care is provided only by one sex. Most, but not all, animals follow the pattern that predominates among human beings: the female usually puts more effort into raising the offspring than the male does.

Since a male ordinarily expends much less energy on a single off-spring than does a female, males potentially can reproduce more often than females. Looked at another way, a male generally has less of its potential reproduction at stake in each copulation. If, for instance, it should mate with a sterile female, it can still have many opportunities to copulate with fertile ones. Females, in contrast, risk much more at each coupling—indeed, in some species the female receives a lifetime supply of sperm at a single mating. If it is from a sterile male, she may have no chance to recoup the loss. Everything else being equal, then, one would expect males to be less discriminating than females. A single mating with a genetically inferior partner might "waste" only a tiny fraction of a male's gametic production. A similar mistake on the part of a female could mean that her chance to pass on genes to posterity is gone.

Based on this sort of reasoning, E. O. Wilson has hypothesized that the basic mating system in animals is polygyny—one male mating with, and normally attempting to monopolize, more than one female. Other systems, monogamy (one male with one female) and polyandry (several males with one female), presumably have evolved from polygyny when ecological pressures mandated it.

But there is work as well as play for males in polygynous mating systems, for they must struggle to monopolize females. That struggle can take many forms. It can consist of a battle royal (a "scramble competition") over females, for example. This sort of thing occurs in some species of frogs, in which the mating process begins when males assemble at ponds and chorus their readiness to mate. Arriving females often are pounced upon by several males at once, with the largest male usually the only one successful at mating.

For many mammals (and, you recall, for cleaner fish), the struggle takes the form of gathering and defending harems; for many other mammals and numerous birds, the trick is for the male to find and defend a rich and therefore (to the female) seductive territory. And males of many other kinds of animals take part in communal sexual displays, advertising their wares (and presumably their fitness) so that females can choose the best among them.

Why should there be such a diversity of methods for organizing sexual reproduction? The answer is that the job is done in many different environments—ones, for instance, with varying distributions of resources. To see how environmental differences could influence the mating system, let's consider a hypothetical species of territorial bird. Suppose that the male birds hold territories in an area where resources (say, insects to eat) are very unevenly distributed. Superior males will be able to hold the best territories (those containing the most insects); in fact, that is what defines the males as superior. They are thus the most desirable mates, for the chances of successfully raising young are higher in a territory of abundant resources. Females naturally should choose to mate with the superior males. And that choice could lead to polygyny.

To see how, consider the plight of an unmated female if all the superior males are already mated. She must either mate with an inferior bird or become the second spouse of one holding a resource-rich territory. In the first case, she has exclusive access to the male and the relatively limited resources of his territory for the rearing of her young. In the second, she must share both the male (who may assist in parenting) and the rich resources with another female. Which strategy is superior would depend on the relative richness of the territories. The point at which the resource distribution is uneven enough to make sharing desirable has been called the polygyny threshold. If the differences are great enough, they might even favor the development of sizable harems on the best territories.

These suppositions seem quite reasonable, but are there any data from nature that indicate whether resource distribution does influence mating systems in this way? It turns out that there are. In the early 1960s, Jared Verner studied populations of long-billed marsh wrens in Washington State. These are small brownish birds with a white stripe over each eye and more white stripes on their backs. As their name implies, they inhabit marshes where they lash their nests to the stems of rushes and other plants. Verner found that bigamous matings (one male mating with two females) occurred even when bachelor male wrens holding territories of their own were available nearby. As expected, male wrens with the territories richest in resources were the ones who acquired additional females. The same pattern has been found in bobolinks and red-winged blackbirds, which, interestingly, also breed in marshes, where the quality of resources may show great spatial differences.

Polygyny is common in mammals as well as birds. One of the most thoroughly studied of polygynous mammals is the yellow-bellied marmot, a large montane rodent closely related to the woodchuck. My longtime friend Kenneth Armitage and his students from the University of Kansas have been engaged for two decades in an exhaustive and productive study of marmot population dynamics and social behavior in the vicinity of the Rocky Mountain Biological Laboratory in Colorado. Over the years, Ken has livetrapped hundreds of marmots, put permanent tags on them, dyed their fur, and released them. He and his students have then spent untold hours watching dyed and tagged marmots —thereby knowing the sex, weight, and general condition of each individual. He has cataloged "amicable behaviors" (greeting, mutual grooming), "sexual behaviors" (genital sniffing, grasping, grappling, mounting), and "agonistic behaviors" (chases, flights, fights, avoidance). These trapping and observation programs have produced a massive directory detailing which marmot is related to which, and who did what and went where. In the laboratory, the Armitage group has also done a variety of studies of marmot physiology and behavior, including a test, using mirrors, of their response to other individuals (which is where Anne and I got the idea for our mirror experiments with reef fishes).

Ken found that male marmots defend territories for four or five months of the year (marmots hibernate the rest of the time). Breeding normally occurs right after hibernation ends in the spring. Later, toward the end of the active season, females shift about, apparently seeking good spots to spend the long high-altitude winter. At that time, there is a second peak of sexual activity. At first it seemed that the quality of the male's territory—especially the hibernating sites within it—played a role in recruiting the harem, just as the quality of a male marsh wren's territory influenced how many females he controlled.

Recently, however, Ken has reviewed more than two decades of data and concluded that his earlier surmise about male territories was incorrect. In fact, the dominant pattern seems to be one of males moving into the most desirable possible areas occupied by groups of related females. The more dominant males are able to take over the best-located harems and defend them against subordinates. Not only does this work show that polygynous systems can result from the operation of different mechanisms (both, as predicted by the theory, related to environmental differences), but also it once again illustrates the great importance of long-term ecological-evolutionary studies that employ a wide variety of approaches. Animal social systems tend both to be complex and to involve subtle interactions. First, and even second, impressions can easily be incorrect.*

Males of some mammals do more than battle other males in scramble competitions or defend territorial boundaries in order to gain access to females. Wandering bachelor groups of male lions, for instance, will attempt to take over prides of lionesses that other groups of males defend as their exclusive property. When a take-over is successful, the new males drive off the large cubs and often kill the small cubs that were sired by the previous group of males. (The females often attempt to defend the cubs, sometimes to the death.) This infanticide appears to be a strategy for quickly making the females, which do not come into heat while caring for young cubs, sexually receptive to the new males. Since a group of males only succeeds in defending its harem for a couple of years on the average, time is of the essence and infanticidal lions are apparently favored by selection over noninfanticidal ones. Similarly, when a dominant male "silverback" mountain gorilla dies, the male who assumes that role in a group often kills infants sired by the last male.†

It is not only among vertebrates, however, that males attempt to monopolize females. In checkerspots and many other butterflies, after

---

* In another interesting example, early impressions of the behavior of red deer were shown to be incorrect by extended studies. Short-term observations indicated that one dominant male did most of the mating and fathered most of the offspring. Long-term observations, however, led to the discovery that, while this was true within a season, a single stag's dominance is brief. Viewed over their entire lives, the disparities in reproduction among male red deer were much smaller—a stag that missed out one year might be top stud the next.
† It has even been speculated that the increase of child abuse in the United States in recent decades (much of which is by boy friends and stepfathers) could be a similar phenomenon, related to the rise in the divorce rate.

A silverback mountain gorilla dining on the vegetation of the Parc National des Volcans in Rwanda. Only a few hundred of these most vegetarian of the great apes survive in the park, whose forests are under constant assault by Rwanda's exploding human population. Each gorilla family group is led by one of these dominant males, who, when they take over a group, often kill infants fathered by the previous silverback. Silverbacks serve as rear guards if the group is attacked by poachers. Despite their fierce appearance and great strength, the silverbacks are easily killed with spears or guns.

"Contact guarding" by a damselfly *(Ischnura gemina)*. After mating, the male (left) continues to grip the female. He will continue to hold on to her neck while she lays her eggs in the aquatic vegetation.

the male has fertilized a female, he deposits a waxy plug in her genital passage, preventing a second insemination (eggs are laid through a separate passage). My student Pat Labine showed by a series of clever experiments that if a female Bay checkerspot was artificially "revirginized" by removal of the plug (but not the sperm) and then mated with a second male, only sperm from the second mating fertilized the eggs. This sperm precedence shows why the males must deposit the plugs—they are protecting their genetic investment. In a sense, the females mated by a single male make up a sort of "harem in time" if not in space.

One of the most thoroughly studied examples of males attempting to monopolize mates occurs in the rather unesthetic arena of mammal droppings. The monopolizers are male dung flies *(Scatophaga),* which hang around in the vicinity of cow pats—they have to, because that's the best place to find females. The dung-fly females, however, only approach the pats when they are ready to lay their eggs on them, so the males generally outnumber females near cow pats about four to one. The males await newly arriving females and, when they approach, mate with them. If no newly arriving female is handy, they may attempt to mate by violently displacing a male that is already copulating or by copulating with a female that is laying eggs.

To prevent the latter, a male dung fly stands guard over a female he has just copulated with while she lays her eggs. This guarding behavior is understandable because in *Scatophaga,* as in *Euphydryas,* there is sperm precedence. Each male thus attempts to protect his genetic investment until the eggs are laid. Male damselflies show an extreme form of this type of guarding behavior, usually remaining physically attached to a female while she is ovipositing. Damselfly males also have special organs on their genitalia that they use to remove a previous male's sperm from the female's reproductive tract—so a male runs a great risk of losing his investment if he does not guard his mate. The behavior of these insects is somewhat reminiscent of a "harem master" bull sea lion, who will guard a bunch of pregnant females on a beach until they have had their pups and he can then have a chance to mate with the females. He, too, is ready to give battle in order to maximize the passage of his genes to the next generation.

Rather than males guarding the female "resource" directly, polygynous mating systems often involve communal courtship displays on a traditional display ground called a lek, within which a male may guard an especially desirable area. The ritualized lekking behavior of grouse and their relatives is familiar to aficionados of nature films. The males assemble on the lek and proceed to strut their stuff, displaying their fancy plumage and inflated air sacs on their necks, making a character-istic drumming call, and dancing with great persistence. Females choose their mates from the displaying males, most frequently choosing the dominant individual who appropriates a prime territory at the center of the lek. The position of a male's territory and thus his opportunities to mate are closely related to his place in the dominance hierarchy. As a result, for example, in black grouse, some three-quarters of matings observed involved only a third of the males. After mating, the females leave the lek to lay their eggs, and the males have nothing further to do with them.

Some male insects, including some fruit flies, display on leks. But not every assemblage of males is necessarily a mating aggregation. Checkerspots seem to have two different strategies for obtaining mates. If there is bare ground, especially on a hilltop, males may perch on it and dash out at passing objects, including other males, females, and individuals of other species. The name of the game, of course, is not to miss virgin females, which are pursued (sometimes by more than one male) and then mated. In such a situation, male-male interactions have a rather stereotyped pattern which appears to be one male defending a perching spot against another. (We have not yet been able to do the critical experiments to see if this, in fact, is the case in checkerspots, but it seems to be in some other butterflies.) In the other strategy, male checkerspots patrol large areas where virgin females may appear and court those they find.

At the montane site described at the beginning of the book, Darryl Wheye and I observed in 1983 a high density of interacting males, but very few females, virgin or nonvirgin, on the road along the ridge. Most of the females were found on the slopes below, where males both perched on the ground in areas of sparse vegetation and patrolled, but in a lower density. This immediately raised the question of why so many males were chasing each other along the road if the virgin females were mostly elsewhere. Was the road actually a lek?

Darryl and I were unwilling to accept as an answer that we had uncovered a group of homosexual butterflies. A more reasonable hy-pothesis was that, on emergence, many females went to the ridgetop to be mated in the same way that female frogs or grouse approach lekking males. Then they might immediately descend to lay their eggs below, where there were fewer males to harass them. Since we rarely observed mating pairs at the site, the question was how to test that hypothesis.

It was Darryl who came up with the brilliantly simple suggestion of dusting the males' genitalia with fluorescent pigment, and that became our major Colorado field experiment for 1984. We marked more than 1,250 individuals. The technique worked like a charm, and it turned out that the males in the hilltop aggregation actually mated only about half as often as the less-concentrated males on the slopes below. Apparently, males that moved upslope to bare areas did have a better chance of mating on average than if they stayed in the relatively dense vegetation at the bottom. But males that perched or patrolled partway up the slope seemed, at least in 1984, to waylay most of the virgins before they reached the vicinity of the males that had gone all the way to the top. In a year in which the population is less dense males that go all the way to the top might have a mating advantage, but in the high density during our study they did not. If the area was a lek, it was a lek occupied by relatively unsuccessful males.

If polygyny is the norm for most animals, under what circumstances might one expect to find its reverse, polyandry, with females keeping a "harem" of males? Although it is the rarest of the three major mating systems, it does indeed occur. Obviously, polyandrous females would have to be emancipated—free to reproduce repeatedly with different mates. Otherwise males would not join the "harem," since no male would gain in fitness by taking up with a female stuck at home rearing someone else's offspring. After all, the idea is to pass on your own genes. (This would not necessarily be true, of course, if the female were a close relative, since, as we've seen, duplicates of an individual's genes can be promoted by helping relatives.)

A few species of birds are polyandrous. In the American jacana, for example, the usual roles of the sexes are almost completely reversed. Males of these raillike wading birds defend small territories in marshes in Latin America. They build the nests, incubate the eggs, and care for the young. The female jacanas, on the other hand, guard large territories that include those of several males. The female helps her consorts guard their territories and may copulate with all of them in a single day. If a clutch of jacana eggs is destroyed, the female quickly replaces them.

This ability to replace lost clutches quickly appears to be the key to the evolution of polyandry in birds. Where the probability of egg loss is high, males may take over the energetically expensive business of rearing the young, leaving the female to put her energy into producing more eggs rapidly when needed. Interestingly, there are recent reports of infanticide in jacanas. Females that take over the territory of another female apparently destroy the clutches of eggs laid by the previous female, even though the males (in the manner of the lionesses) attempt to defend them. The female then copulates with each male and provides her own eggs for him to rear. If confirmed, this would be the only known example of infanticide by females as a regular feature of a mating system —but it is one that makes evolutionary sense, since it would promote the perpetuation of the genes of the female that won the territory.

Kenneth McKaye, a productive young ecologist who directs a field station on the shores of Lake Malawi, has described certain cichlid fishes from that African lake in which polyandry and polygyny are combined. The males set up leks on the sandy lake bottom, each individual occupying a territory on the lek in which he builds a nest where mating and egg laying occur. Females hover above the nests and choose their mates from among the available males, descending to spawn in the nests of the chosen ones. After the eggs are laid and fertilized, the female takes them into her mouth and carries them until they hatch. As with most lek systems, a male may be visited by many females. What is unusual about this system, and what makes it both polygyny and polyandry, is that a female may mate with several males—it might lay only a single egg with any individual male. The evolutionary significance of this system has not been investigated in detail, but one would expect it to result in less of a concentration of reproductive success in a few males than in the more typical lek system of the grouse.

While polyandry is occasionally found in birds, mammals are almost never polyandrous. This is not surprising, considering the physiological need for the female to nurse the young. In mammals, females normally cannot be emancipated from the care of offspring.

Even though polygyny may be the fundamental mating system among animals in general, something over 90 percent of bird species appear to be monogamous—a single male mating with a single female for a lifetime or during a season (or at least during a single nesting in birds that nest more than once per season). The same habit is found in various mammalian groups, although it is not nearly as common in mammals as it is in birds. The principal reason for the evolution of monogamy in birds seems obvious; in most cases the young have a reasonable chance to survive only if two parents share the tasks of rearing them. This view is supported by the observation that both polygyny and polyandry in birds are found predominantly in species such as grouse and ptarmigan, whose chicks are able to move about and search for food on their own as soon as they hatch. The species of birds whose young hatch still blind and helpless are the stronghold of monogamy.

In mammals, monogamy tends to occur in territorial species in which the variation in resources from one male's territory to another's is minimal. In these circumstances, the benefits to a female of mating with (and sharing the territory of) an already-mated superior male are less than those of taking up with a single (slightly) inferior male. In other words, the polygyny threshold is not crossed. The mechanisms by which monogamy is maintained vary. To take but one example, in some gibbons it is enforced by already-mated females who prevent other females from mating with their consorts. In this case, resources seem to be too scarce to support a polygynous system—that is, a territory might not be able to support more than two adult animals and their offspring.

Before closing our discussion of animal mating systems, I should mention cooperative breeding. In a number of bird species, more than two adults provide care for a given offspring. In some cases, as in the Florida scrub jay, nonbreeding adults help rear the young of those breeding. In others, such as the groove-billed ani of tropical America, several pairs of reproductive adults share the rearing of all their young.

Certain mammals also engage in cooperative breeding. For instance, Patricia Moehlman of Yale University has done a fine study of black-backed jackals on the Serengeti Plain of Tanzania. She found that frequently there were "helpers at the den"—adults that helped breeding pairs raise their young by both feeding and guarding them. Often the helpers were pups of the previous year's litter, older brothers and sisters of the pups they helped to raise.

When adult birds or mammals forgo reproducing themselves to help others, the question of the evolutionary origin of the behavior arises. Often the answer can be found in inclusive fitness—an inexperienced young adult may be able to promote copies of its genes more effectively by helping to raise siblings than by attempting to breed. Other possible explanations include gaining experience and avoiding the physiological strain and increased risk of predation involved in breeding when inexperience makes success problematical.

The complexities of sexual relationships in animals are gradually being unraveled, but sexual systems in plants are, if anything, more puzzling. Some plant species, like some animals, reproduce asexually. Anyone who has had an aquarium containing *Vallisneria* has seen this —this grasslike aquatic plant sends out runners from the base of its leaves, and at points along these runners, new plants develop.

Other plants are hermaphroditic, each flower being *perfect*—containing both male* structures that produce pollen and female structures that bear ovules. In some cases, pollen can fertilize ovules on the same plant; others have devices to prevent self-fertilization. Some plants produce two kinds of flowers, male and female, with both kinds found on the same individual. In others, the condition that we human beings look upon as "normal" exists: male flowers and female flowers are found on different individuals.

The complexity doesn't end there, however. Some plant species bear either male or perfect flowers but no female flowers; others bear either female or perfect flowers but no male flowers. And some, such as the jack-in-the-pulpit, reverse the cleaner-fish sequence and function as males when small and as females when they grow larger. Furthermore, female jack-in-the-pulpits that shrink due to lack of light or nutrients can switch back to functioning as males.

Much of the discussion about the reasons evolution has produced such diversity in reproductive strategies in plants has focused, as did the discussion of why sex itself evolved, on the amount of recombination produced, although other factors clearly play a role. Many plants, faced with the problem of getting their gametes together, use wind or water to carry pollen grains (tiny haploid male plants that produce sperm) from one diploid plant to another. But many other plants have produced flowers, nectar, scents, and other devices to get animals to transfer their pollen. For example, some plants that are pollinated by bats have male and female flowers, usually segregated onto separate individuals—but occasionally these plants bear perfect flowers. The bats are attracted to certain floral parts, which they eat, and there is less damage to female structures when the flowers are separate than when they are perfect. In

---

* Some botanists object to the use of the terms "male" and "female" for floral parts since technically the structures produce not male and female gametes (as do male and female organs in animals) but tiny male and female plants—pollen grains are male plants, and structures within the ovule of a flower are female plants—which in turn produce the gametes.

this case, considerations of recombination seem secondary to avoidance of being destroyed in the process of pollination.

Perhaps the greatest mystery in the world of plant sex is presented by that commonest of flowers, the dandelion. Many dandelions reproduce entirely asexually, and yet they have fully formed yellow flowers that manufacture nectar and pollen. Such attractive flowers and nectar are clearly devices that evolved to con insects and other animals into helping plants with their sexual activities—and a wonderful con it is, being the basis for the great diversity and success of the flowering plants. But why, if it reproduces asexually, does the dandelion spend so much energy on sexual attributes?

Could it be that the production of flowers and nectar is so ingrained in the developmental system of dandelions that they have thus far been evolutionarily unable to reorganize themselves to skip the manufacture of yellow petals and nectar and save the energy now wasted? A weak explanation, but it will have to do until someone thinks of a better reason for them to lure and provision unneeded pollinators. (There is, however, no question that in many cases the evolution of organisms is constrained by structural commitments made in the past. To take an extreme case, no amount of selection pressure in favor of flight is likely to get hippopotamuses to take to the air.)

To me, mating systems are among the most intriguing of all ecological topics—their great variety underlining both the enormous diversity of environmental situations to which organisms must respond and the extreme evolutionary flexibility of those responses. But now I'd like to move on to look at some other aspects of sociality—starting with a group of animals that is very effective in helping plants with their sexual problems: bees. Bees, especially the honeybees, are also the best known of the social insects, a group that includes ants and various kinds of wasps and bees (all very closely related to one another) and termites (not closely related to bees, but instead close relatives of cockroaches).

Many volumes have been written on social insects (I strongly recommend Charles Michener's *The Social Behavior of the Bees* [Cambridge, Mass.: Belknap, 1974] and E. O. Wilson's *The Insect Societies* [Cambridge, Mass.: Belknap, 1971] for a thorough introduction to these endlessly interesting insects), and I will not attempt to detail insect social behavior here. Instead I want to look at another aspect of sociality that is as interesting to ecologists as the evolution of mating systems: how and why did the fascinating and advanced patterns of insect social behavior evolve? What environmental conditions have produced these complex societies, and why haven't all major groups of insects developed truly social species? As you will see, we do not yet have comprehensive answers to these questions.

A basic feature of the most highly regimented societies of insects is a division of labor based on castes—behaviorally (and often physically) differentiated groups of individuals specialized for different tasks. For example, the "busy bee" that you observe hustling from flower to flower is a sterile female worker, slaving to gather nectar and pollen to improve the reproductive success of her mother, the queen bee. To find a reasonable explanation for her behavior, we must turn to the unusual genetic sex-determination mechanism of the Hymenoptera (the insect order that includes wasps, ants, and bees) and to the idea of inclusive fitness presented earlier.

In the Hymenoptera, fertilized eggs produce females and unfertilized eggs produce males—a system that goes by the ten-dollar name of haplodiploidy, since the males are haploid (have only one set of chromosomes received from the mother) and the females diploid (have two sets, one received from the mother and one from the father). This leads to the peculiar result that, in wasps, ants, and bees, sisters are more closely related to one another than mothers are to daughters. How can this be? First consider that a haploid father passes the same sets of genes to each of his daughters, so that the paternal inheritances of those sisters are identical. A daughter receives half of her chromosomes (and genes) from her mother, so their relatedness is one-half (since both mother and daughter are diploid and do not share the genes on half their chromosomes). In the maternal half of their inheritance, each gene of two sisters has a 50:50 chance of being identical—that is the chance they will receive a copy of the same member of the pair of chromosomes that in their mother carries that gene. Since the paternal half of the sisters' inheritance is identical (100 percent the same), and the maternal half is 50 percent the same, on the average three-quarters of the genes in two sister bees are identical. Another way of looking at this relatedness of three-fourths is that half comes through their father and a quarter through their mother—which is the essence of this complicated matter.

Thus, because of the haplodiploid sex-determination system, young female Hymenoptera can, in theory, increase their inclusive fitness more by helping to raise their sisters (who are carriers of copies of three-fourths of their genes) than by having offspring of their own (who would have copies of only half of their genes). Haplodiploidy, which is universal in the group, may explain why social behavior arose separately in the several different groups within the Hymenoptera. Haplodiploidy, it would seem, favors the evolution of social structure.

These sorts of inclusive-fitness arguments, while not universally accepted, have provided a way of thinking about the evolution of insect societies that most ecologists find more plausible than those that have gone before. Major questions, however, remain to be answered. Why, for example, if haplodiploidy predisposes the Hymenoptera to form societies, are there so many solitary (nonsocial) Hymenoptera? What ecological factors tip the selective balance toward sociality? How is the transition from solitary to social system made?

Partial answers to these questions are emerging from extensive research on bees by Charles Michener and his colleagues. They have found a spectrum of social behavior ranging from unrelated females nesting communally, through partial dominance of some females by others, to the queen-and-sterile-female-workers system in the most highly social bees. A logical conclusion to draw from such a contemporary sequence is that it reflects different stages that must have occurred in the evolution of bee societies. If Hamilton's ideas on inclusive fitness provide one answer to the "why" of the evolution of hymenopteran societies, Michener's findings give us a clue to the "how." And researchers are now vigorously searching for more complete explanations.

Even more puzzling than bee sociality, however, is the evolution of complex termite societies. No haplodiploidy to lean on here; both sexes hatch from fertilized eggs, and the relatedness of siblings is one-half, regardless of sex—just as in butterflies, mole rats, and people. In detail, termite societies are very different from those of bees, wasps, and ants. The ecological pressures that gave rise to complex societies in the Hymenoptera and the termites (to say nothing of the naked mole rats) were probably also very different.   .

Indeed, the basis for the evolution of sociality in termites may be intimately connected with their most widely known attribute—the ability to eat wood. Along with certain primitive roaches, they are the only animals able to accomplish this feat. They do it with the aid of certain protozoa (single-celled organisms) that live in their guts; the termites and roaches eat the wood, but the protozoans digest it. These protozoan cultures necessary for their survival are passed from generation to generation by the (from our prejudiced viewpoint) less-than-fastidious habit of the young eating from the anuses of their elders. This necessitates an association of generations that is unusual for insects—at least a minimum level of sociality, which may have provided the selective springboard that led eventually to termite high society.

Sex and society in nonhuman organisms is a fascinating and complex subject, as is sex and society in *Homo sapiens*. But can the sorts of reasoning that have been applied to the ecology and evolution of social behavior in other animals be applied with profit to human society? Many biologists, including myself, think so—although to what degree and to what ends remain highly controversial.

Trying to put any aspect of human behavior into an evolutionary perspective carries with it the risk of becoming embroiled in some version of the antique "nature-nurture" controversy. On one side (to exaggerate), there is the biological-imperative, man-as-ape view that our genes condemn us to be forever aggressive (or territorial, or jealous, or superior to other races or kinds of people, or whatever). On the other side is the *tabula rasa,* everybody-would-behave-identically-were-it-not-for-environmental-differences position, which implies that while four billion years of evolution has led to genetic influences on the behavior of all other organisms (and all other features of *Homo sapiens*), it somehow has (miraculously, it would seem) left human behavior unscathed.

Worse yet, these two equally ridiculous views tend to be associated with political persuasions. Push the button of a right-wing racist and you'll likely get a speech on the bad genes that produce stupidity in blacks or avariciousness in Jews. Go to lunch with an ultraradical feminist and she may explain to you how social conditioning is the only thing that makes men behave differently from women; hormones have nothing to do with it. Taking a prominent position—any position—on the evolution of human behavior is somewhat like standing up in no-man's-land during the battle for Verdun, as E. O. Wilson discovered to his dismay after his monumental book *Sociobiology* was published.

Before turning explicitly to human sexuality, however, let us take a moment to consider human behavior in general. As you might guess from the discussion so far, the nature-nurture dichotomy is a false one. Genetic endowments are expressed differently in different environments; development is basically a consequence of gene-environment interactions. Thus possession of certain genes does not necessarily preordain anything. All the "obesity genes" in the world won't make a starving child in Ethiopia fat, nor will an Einsteinian genotype necessarily turn that child into a nuclear physicist. On the other hand, there is no reason to think that the best of all growing environments would have made Mickey Rooney taller than Wilt Chamberlain or that the best of all learning environments would have made Adolf Hitler as eloquent a writer as Richard Wright or as brilliant a physicist as Albert Einstein. Changes in genetic makeup or in environment can alter the gene-environment interaction and result in dramatic changes in the phenotype—but the potential for change is not without limit.

One would expect certain inherited characteristics of human beings to be *relatively* buffered from environmental changes. The genes that control the overall patterns of physical development produce the same general kinds of results in a wide variety of environments—if they did not, selection would quickly correct it. People with legs growing out of their ears, eyes on the soles of their feet, or lungs connected to their intestines probably would not pass many of their genes to future generations. Such things as age at puberty and adult size *are* sensitive to the nutritional environment, but the arrangement of body parts, physiological functioning, and so on turn out to be essentially the same in all human populations. No environment regularly induces the sort of arrangements I just listed, because if selection could not buffer the developmental system to prevent it, human populations could not survive in that environment.

This resistance to environmentally induced change in basic physical development makes obvious evolutionary sense. But such insensitivity to the environment would not be selected for in many behavioral and intellectual traits. Flexibility of response to the environment in such characteristics is one of the hallmarks of human success. And that flexibility is best seen in the human capacity to develop, elaborate, and modify the large body of nongenetic information that is called culture, a capacity that sets us apart from the rest of organic creation. Genetic evolution is always slow, with populations changing over time measured in generations. Cultural evolution can be fast. Nothing resembling the portable computer on which I wrote this book existed less than half a generation ago, and the particular model I used was designed and produced in the last three years.

But selection should not produce a human brain that is a true *tabula rasa,* a computer that the environment can program entirely without restraint. The ability to be programmed is important, but it automatically entails the ability to be programmed wrong—to be fooled or taken advantage of, perhaps with evolutionarily lethal results. To take an extreme case of the latter, suppose the "purely cultural" society evolved toward homosexuality as the exclusive form of sexual behavior. Selection favoring individuals even slightly biased by their genetic makeup toward heterosexuality would surely be strong, since strict homosexuals do not reproduce. Clearly, though, the tendency to have what might be called "genetic biases" or "genetic predispositions" should be strongest for those cultural characteristics most directly affecting reproductive success. Choice of sex of mate should be more highly "programmed" than the choice of a major in college.

Equally clearly, genes are not destiny. The sensitivity of human behavior to the environmental (including cultural) milieu is extraordinary. The vast diversity of human cultures, behaviors, and ideas is evidence of this. So are certain behaviors directly relating to reproduction and parenting (and thus fitness). After all, rather than routinely killing the offspring of previous mates when taking up with a new female, as lions and gorillas do, *most* human males love and nurture them. Similarly, couples commonly adopt and rear children who share none of their genes, and many people voluntarily forgo reproduction, often having operations that make it impossible. Presumably none of these behaviors would occur if our genes were in complete control.

Against this background, what can be said about the mating system of *Homo sapiens*? The first thing is that there is no sharply defined estrus period, a situation unique among mammals. Females of other mammals have a limited (and advertised) period of female receptivity, or "heat" —the time when they are available for copulation because they are ready to conceive. Human females, in contrast, can become aroused at almost any time, and they not only conceal their fertilizability from males, but to a large degree from themselves. Thus the old joke that the rhythm method of birth control is called "Vatican roulette" and its practitioners most often called "parents."

The year-round sexuality of *Homo sapiens* has results that pervade all human societies and produces for many of us one of life's greatest pleasures. But its origins remain quite mysterious. The enormous plasticity of human behavior makes it very difficult to evaluate genetic variation in human sexual conduct—and, since scientists are either men or women, a sexual bias may well influence the judgments of investigators.

Furthermore, virtually nothing is known of the ecological situation in which the basic human mating system evolved. All we know is that it evolved in the last few million years after the human lineage diverged from those of the other higher primates. We know this because all of the great apes have a clearly defined estrus period.

It is therefore understandable that, lacking information on both genes and environment, scientists have fallen back on speculation to explain continuous sexuality—the central mystery of human sexual behavior. The most common speculation among biologists has been that the loss of estrus was a device that evolved to keep the man associated with the woman through the very long period in which the human child is dependent. But, as Donald Symons explains in his provocative book *The Evolution of Human Sexuality* (Oxford University Press, 1979), there are severe defects with this view. And they come from both sides of the gene/environment fence. For example, gibbon couples form life-long monogamous bonds, but they do not enjoy the pleasures of continuous sexuality. Indeed, examination of the cultural complexities of marriage in many human societies shows little evidence that the primary cement in the pair bond is sex. Rather, marriage most often seems to be based on economic and social considerations, not lust. There is certainly some evidence for the view, attributed to Alfred Kinsey, that marriages that succeed do so "not because but in spite of the sexual obligations . . . entailed."

Symons himself offers two speculations on the loss of estrus, based heavily on parallels between human and chimpanzee societies. One is that being continuously "copulable" helped females to beg meat from successful males after hunting became the primary male activity. The benefits of superior nutrition presumably overcame the costs of remaining permanently sexually active. The second speculation is that lack of estrus is a woman's "device" that evolved *after* the development of marriage. It is a device to make it more difficult for a husband to control his wife's sexual activities, since he must guard her always to be sure he is not cuckolded, not just (as in other primates) when she is in estrus. Thus it is a device that increases the possibility of the woman selecting a different biological father for one of her children—the strong, handsome brute instead of the rich old man selected by society.

Symons comments: "I have little confidence in these scenarios; probably both are wrong, but they need not be right to be useful or interesting, they only need to be as good as or better than, competing explanations." These scenarios—like the explanation of the woman's sexual receptivity causing the man to hang around while the kids need care—all focus attention on forces that increase the reproductive success of the woman and thus are evolutionarily plausible.

The search for such selective forces has refocused attention on the old question of the degree to which differing male and female attitudes about sex can be traced to cultural influences or biological evolution. Symons argues for a substantial influence of the latter, based on the presumed influence of the asymmetrical parental investment of the two sexes. Men are able to promote their genes by mating with a large number of women. Even though individual offspring may not receive the best of care (since each receives, on the average, less paternal care —or none at all), their large numbers can compensate for a lower survival rate. Women, by contrast, can gain relatively little by being promiscuous, since they are physically limited in the number of children they can bear. Symons says that this has led to a fundamental sexual difference in attitude toward numbers of partners: a man tending to seek numerous copulations (to maximize the number of children sired); a woman seeking a stable relationship with a single good provider (to maximize the chances of survival of her limited number of children). Furthermore, the desire of mature men for younger partners Symons claims to be related to the greater success of younger women at child-bearing. On the other hand, females might prefer older men with established track records as providers.

Just because something makes a good evolutionary story, however, does not mean that it happened that way—as Symons himself says. He has assembled an impressive body of evidence for his views, however. Among the most persuasive to me is evidence of male-female differences in ease of arousal and desire for sexual variety. In addition, this difference can be reduced hormonally. Females suffering from adrenogenital syndrome, in which an excess of the male hormone (androgen) is produced, tend to have masculine patterns of arousal (such as stronger reactions to visual stimuli). These sexual differences in arousal patterns *could* be explained by a convoluted hypothesis involving only environmental factors. But the notion that they are largely a result of genetic differences between the sexes (hormonally conditioned—with hormones in turn largely genetically controlled) seems considerably more parsimonious.

So what if Symons is (as I suspect) largely correct that human evolutionary history has given males and females different genetic predispositions for some traits (such as desire for sexual variety)? It is a conclusion that many people find very threatening because they mistakenly believe that genes predetermine rather than predispose. In the face of massive evidence to the contrary, they are afraid that husbands will necessarily be unfaithful, or that women will always do better as mothers than as executives. Or they may fear that if males and females show hereditary behavioral differences, there might be other psychological differences that are dictated by heredity—that, for example, all children of parents with low IQs also will be so afflicted.

To believe these things, however, is to be ignorant of a basic fact of biology in general and evolutionary ecology in particular. To repeat: The environment shapes the genetic endowment of individuals through the medium of natural selection, and, in addition, it helps to determine their phenotypes by interacting with their genotypes in the course of development. To assume, *a priori,* that human behavior has not been influenced by biological evolution is ridiculous; to assume our behavior to be rigidly constrained by that evolution is equally ridiculous.

Sociobiology is the discipline that attempts to integrate an ecological-evolutionary perspective into the social-science view of "human nature," as in Symons' work mentioned above. Indeed, much of the material in this chapter can be considered sociobiological. While sociobiology has done much to increase our understanding of sex, societies, and so on, it has also triggered a gigantic controversy—a new version of the nature-nurture one—that has been thoroughly aired in everything from tabloid newspapers to the technical literature. In closing this chapter, I would like to look at one issue that has been raised in the controversy which is of particular importance to ecology and ecologists.

The claim has been made that the efforts of Wilson and others to look at genetic influences on behavior should be condemned because the results could give aid and encouragement to fascists and racists. The results, for example, might be used to justify forced sterilization of people with low IQ scores or the "wrong" accent, in order to remove their "bad genes" from the population. Although I believe that no legitimate interpretation of sociobiological theory would justify such actions, the fear is well founded, since zealots of all political stripes have never hesitated to misinterpret science in the service of their cause. The broader question then becomes: Should scientists investigate areas or announce conclusions that might support a certain political view or be misinterpreted as supporting it? Virtually all ecologists are aware of and concerned about human overpopulation of Earth. Should they (as I have been told with some frequency) refrain from saying that there are too many people, because someone might interpret that as a reason for killing off a surplus?

Unfortunately, there are few areas of science that don't produce conclusions, discoveries, or devices with sociopolitical implications. And predicting where such implications are likely to arise is not simple —after all, at one time nuclear physics seemed among the most esoteric, least applicable of sciences. One cannot yet judge even whether the entire enterprise of science will prove a net benefit to *Homo sapiens*. If humanity perishes in a nuclear winter, it clearly will not have been. But it is certain that pure science will continue to be a major avenue by which human beings increase their knowledge and that attempts to restrict the search or the dissemination of the results will mostly prove, in the long run, fruitless.

In my view then, the best course for ecologists and other scientists to follow is to pay a lot more attention to the social consequences of their activities and, at the same time, keep the public and decision makers informed about scientific progress. Decisions about which lines of inquiry to pursue are already in part made outside of the scientific arena by corporate executives, legislators, the military, and others. It behooves all of us to make sure these decision makers know something about what they are doing. The future will bring increasingly difficult questions for scientists and society to answer, in fields ranging from weaponry and genetic manipulation to ecosystem management and preservation.

Above all, the fates of the scientific enterprise and *Homo sapiens* now seem to be inextricably connected. If both survive and prosper, I believe it will be for two reasons. One will be that accurate information about science, what it has discovered, and how it works, has become much more widely disseminated. The second will be that more and more people will learn there are other sources of knowledge besides science—that, for example, science cannot provide sure solutions to many ethical problems. It can often clearly delimit ethical alternatives, but the wisdom to select the right one will, in many cases, have to be found elsewhere.

ONE-ON-ONE

Predation, Mutualism, and Competition

THE WORD "predator" produces a little tingle in the spine of most people. Perhaps it conjures up a vision of that giant shark stalking a naked woman on a midnight swim in the opening scene of *Jaws*—or some equally harrowing scene. Encounters with large predators are often exciting for ecologists too. They make up some of my most vivid memories from four decades of field biology. For instance, one bright subarctic evening in 1955, some butterfly-collecting friends and I were slowly driving along a road in Mount McKinley National Park in Alaska, hoping to get pictures of a grizzly bear, the largest land-dwelling predator in North America. It was late spring, and the park was almost devoid of tourists. We allowed a pickup truck to pass us, and then suddenly we saw, right ahead of the truck, a big Toklat grizzly sow and her yearling cub.

Rather than stop, the pickup immediately charged the bears and then pursued them down the road, enraging us and, I suspect, the mother bear. After being chased about a hundred yards, the bears split, the sow going upslope and the cub downslope. The driver of the truck immediately slammed on his brakes, and his companion leaped out and (for no reason apparent to us) ran waving his arms at the cub, which had stopped about ten yards from the road. I readied my camera as the sow observed the action from about ten yards on the other side of the road—I figured that a picture or two of a fool being dismembered by a mother grizzly would have some commercial value. No such luck, however. The sow didn't charge, and after a few awkward moments the cub simply circled around the waving man to rejoin its mother, and they departed.

Bears and sharks are unpredictable; both on occasion eat people, but most encounters end with the nonhuman predator beating a retreat in the face of the most dangerous animal on the planet. In fact, the only prey I've observed a grizzly eat was a mouse that a large male bear spent a half hour digging out of a tundra slope in McKinley Park. He couldn't possibly have recovered his energetic investment.

Predator-prey interactions are one general set of relationships between ecologically intimate organisms. Defined broadly they include parasite-host and herbivore-plant, as well as the classic carnivore-prey, interactions—all those relationships that involve one kind of organism eating another, even if the eater is tiny compared with the eatee. Besides predation, there are two other types of relationships that involve basically one-on-one interactions (which we will look at later in the chapter). These are mutualism, where two species benefit each other, and competition, where one species is harmed by another (or both are harmed) without either organism eating the other. These kinds of interactions between species whose activities are closely intertwined ecologically are the starting point of community ecology. Understanding them is basic to solving the puzzle of what controls the mix of organisms found living together in any given area.

Everybody (except green plants) has to eat somebody else, so in a way predation would seem the most fundamental of processes. When we looked at the acquisition of nutrients in Chapter 1, the focus was on individuals. Indeed, a burgeoning literature on "foraging strategies" deals with questions such as how selective individual predators are and under what circumstances they will shift from one prey species to another. But in this chapter, our attention focuses not on the individual— whether a foraging grizzly "makes a profit" on a single hunt, for instance—but on populations. And, as usual, it is here that things get complicated. In considering groups of individuals, ecologists are confronted with many new questions, such as: Can a population of predators and a population of prey coexist? If they do, how will their population sizes be related? Do predators usually control the size of populations of prey? What happens to the predator population if the prey population gets smaller? Or larger? Have predators evolved "prudence" to prevent them from overeating their prey?

These questions are of much more than theoretical interest. Human beings often operate as predators—for example, as hunters of oceanic fishes. Understanding predator-prey interactions is important to designing sustainable harvesting strategies. And people are frequently prey to a large number of organisms. Indeed, epidemiologists (who study the course of diseases in populations) are paying more and more attention to the ecological and evolutionary relationships of *Homo sapiens* to its smaller predators, from intestinal worms and the protozoan parasites that cause malaria to bacteria and viruses. In addition, pests of crops and carriers of disease can often be controlled by predators—either naturally or with human help. Knowing how and when to assist such beneficial predators usually requires detailed knowledge of the interactions between them and their prey.

One of the first investigations of predator-prey relationships was carried out by pioneer Russian ecologist G. F. Gause. In the early 1930s, Gause studied populations of two protozoan species in the laboratory. One, a member of the genus *Paramecium,* was the prey; the other, a *Didinium,* the predator. When the two were placed together in a test tube of water containing food for *Paramecium,* the results never varied. The *Didinium* found and ate all the *Paramecium.* Then the *Didinium* themselves starved to death. The predator population destroyed the prey population and then went extinct.

This probably was not a pleasing result for Gause, since laboratory experiments (like theory) are supposed to help one to understand nature by simplifying it—not by producing results contrary to all experience in the real world. In nature, predators do not normally eat all the prey and then go extinct. Robins eat worms, but every year there are still robins and worms. Back to the drawing board.

Gause next tried creating a refuge for the *Paramecium* by putting some glass wool in the bottom of the test tubes. This didn't work out too well either. Some of the *Paramecium* were indeed able to escape from the *Didinium,* but then the *Didinium* starved after devouring the accessible *Paramecium* and the hidden *Paramecium* subsequently re-populated the tube. The predator population went extinct and the prey population recovered. Gause finally was able to get the two protozoans to coexist by simulating migration from an outside source—every once in a while he introduced more *Paramecium* and *Didinium* to the system, simulating a heterogeneous environment that included both the original setup and an "island" that periodically dispatched migrants.

Mighty close to cheating, you say? Perhaps, but nature contains precious few genuinely isolated systems, very few in which a predator species has only one prey and vice versa, and very few in which habitat diversity is nearly zero. Attempts to get a predator and prey to coexist in a microcosm (literally "little universes" designed to represent the real world) have been successful for even limited periods only when the microcosm is large enough to contain a great deal of structural hetero-geneity—sites favorable to the organisms separated by barriers. Then migration between the sites prevents the extinction of either species within the microcosm as a whole.

Clearly, in a simple system, predators can make a dramatic impact on prey populations and prey populations can control the destiny of the predators. Is there any evidence that they can do so in nature? The most famous putative example of such strong predator-prey interactions in nature was drawn from the fur trade records of the Hudson Bay Com-pany. The number of furs of lynxes and snowshoe hares sold to the company annually has been recorded since early in the last century and is considered a reliable index of the size of the populations of these animals. The records show that population sizes of the two species fluctuated, with peaks of abundance about every ten years. The lynxes, which eat the hares, generally have large populations at the same time or just after the hares do—that is, the peaks of the population curves of the two species more or less coincide, with those of the lynxes tending to be slightly after those of the hares.

At first glance, this pattern seems to make sense. When hares are abundant, the well-fed lynxes easily raise their kittens and the lynx populations swell. The greater numbers of lynxes then decimate the hare populations, leading in turn to hard times and a population decline for the lynxes. Relieved of the predation pressure of the lynxes, the hare populations build again, and the oscillations continue. All very nice, but quite probably at least partly wrong. The lynxes can't breed, and their kittens can't mature, fast enough to overtake and suppress the hares so quickly. Furthermore, on some islands from which the lynxes are absent, the hare populations fluctuate just the same.

The reasons for the cycles of these two species probably lie in the interactions of other factors with the lynx-hare predation system, especially the relationship of the hares to the plants they eat. For example, a decline in available food or a change in its quality may help to slow the hare population explosions. One theory proposes that lack of food ends the hare outbreaks, then the lynxes and other predators reduce the hare populations to very low densities. And, it has been suggested, it may be the hares that assault the lynx, and not vice versa, by being carriers of a lynx disease.

Canadian scientists, under the leadership of ecologist Charles Krebs, are launching a large-scale investigation of this perplexing problem. Among other things, Krebs and his colleagues have been hauling tons of laboratory rabbit feed onto experimental plots during frigid Yukon winters. The feed supplements the hares' diets, thereby testing the possible role of food shortages in causing hare populations to collapse. With luck, these ambitious investigations will help to solve the puzzle.

Regardless of what drives the "coupled" hare and lynx cycles, laboratory systems like Gause's, in which refuges and migration prevent extinction of either predator or prey, show that interactions of predators and prey can cause such coupled cycles. Mathematical models (which can be formulated to exclude extinction of either predator or prey) of predators and prey interacting only with each other can also produce coupled cycles. Here again, a model can be helpful in thinking about the real world.

The granddaddy of predator-prey models is the Volterra model. Even without describing it, I can give one of the model's most interesting and nonintuitive results. If predator and prey populations are more or less in balance with one another (which they would *not* be if, for example, a few predators had just invaded a gigantic population of prey), and if an environmental change raises the death rates in both predator and prey populations, there will be a disproportionate decline in the number of predators.

This result is known as the Volterra principle. It predicts what will happen, all else being equal, after an application of a broad-spectrum pesticide (one that kills many species of insects) to a farmer's field in which herbivorous insect pests are being eaten by predatory insects. The surviving prey, whose death rate in the aftermath of the pesticide is now lower due to the absence of predators, will suffer less than the surviving predators, which cannot find sufficient prey to increase their birthrate quickly. The population of prey (the pests) will thus recover from the pesticide much more rapidly than the predators and build up a larger population than was previously present. Thus an even-handed assault on a predator-prey system will tend to promote the prey and suppress the predators—in this case, making the pest problem worse. This, of course, is exactly the opposite of the intended result but exactly what happens all too often from the use of pesticides, not only because of the Volterra principle, but also because of several other factors that will be discussed later.

In the simple Volterra predator-prey model, as in simple laboratory experiments, the systems are not very stable. Even if the simplest model is made more realistic by incorporating a time lag between the increase in the prey population and that of the predator (biologically a very reasonable assumption—predators don't reproduce instantaneously), either the predator or the prey or both eventually go extinct. If, however, further complexity is added, and the models are thus made more biologically realistic, by considering such things as hiding places for the prey (so that proportionately more individuals have good hiding places when the population is small than when it is large), limits on food available to the prey, territoriality of the predators, and so forth, the modeled system becomes more stable, more like the long-term coexistence seen in nature.

Mathematical models have suggested other interesting influences on the stability of predator-prey systems. One is the so-called "paradox of enrichment." It turns out that the carrying capacity of the environment supporting the prey species is often important to stability. If that capacity is increased and more resources become available to the prey, the system may become unstable. Destabilization can occur when populations of phytoplankton in a lake are more or less in equilibrium with the insects and other predators that feed on them and fertilizer runoff from farms suddenly fertilizes the lake. The additional nutrients may cause a bloom (an enormous increase) in the plankton population if its growth were previously limited by lack of nutrients. The plankton bloom, in turn, shifts the system away from equilibrium, leading to oscillations in which some prey and some predator populations go extinct. Enriching the system, paradoxically, may thus do it great harm—by the time that the plankton exhaust the nutrients and their bloom ends, the system may have been considerably simplified.

While predator-prey systems can be studied fairly easily with mathematical models and, in some organisms, in laboratory microcosms, they frequently are difficult to study in nature. One of the greatest problems is that the act of predation is not often observed. In addition to the grizzly's capture of the mouse, I've only seen one other successful attack by a large predator. This occurred in the Lemaire Channel off the coast of Antarctica, where Anne and I watched a pod of killer whales knock a Weddell seal off a pan of ice and devour it. I've seen only a few misses too. An African lioness once unsuccessfully used our car as cover to stalk a wildebeest. In McKinley Park, I saw a wolf rush a herd of Dall sheep. The wolf probably was just testing to see if any were sick or injured; the sheep evaded its charge with ridiculous ease. Small predators, however, are easier to see in action—spiders netting flies, mirid bugs spearing newly hatched checkerspot butterfly caterpillars and sucking them dry, warblers gleaning insects from tree limbs, even an occasional fish being gobbled by a larger fish on a coral reef *—but still relatively rarely observed.

Thus, even though predation is common, observing it systematically usually requires either luck or dedication. George Schaller studying Serengeti lions and Hans Kruuk following the lives of the hyenas of the Ngorongoro Crater both had dedication. Once Anne and I had some luck. We were on a short trip to southwestern Brazil, attempting to do research on the interaction of *Euptychia* butterflies and their host plants (as we had at Simla). The research had not been going well, primarily because we were having trouble finding undisturbed habitats. Our Brazilian hosts had remembered relatively undisturbed forest in the vicinity of Iguaçu National Park. We, of course, wanted to see the famous Iguaçu Falls, one of the world's great cataracts. So there we went. But in the few years since our friends had last been to Iguaçu, the forest had been cleared not only to the park boundary but in some places into the park itself. Indeed, what we saw of the state of Paraná looked like the

---

* Of course, the *importance* of predation often must be inferred even when it can be observed. For example, importation of a predatory beetle and a predatory fly effectively controlled a scale-insect infestation that was ruining California's citrus industry. Individual predators could be observed at work, but their importance was seen in dramatic declines in the abundance of the scale insects.

Midwest of the United States, its rich soils now supporting corn and soybeans.

We went into the park, where butterfly collecting is prohibited, as it is in U.S. national parks, to see the falls. Butterflies were extremely abundant, for the river itself and the clearing for walkways and viewpoints had created exactly the kind of forest-edge environment in which they and their food plants thrive. At one point we stopped to observe a group of medium-sized lizards (four to six inches long, not including tails) on rocks adjacent to a walkway near the brink of the falls. Occasionally, spray from the falls drifted over them. As we watched, a small, colorful "89" butterfly (a species of the genus *Callicore,* so named because the pattern on the underside of its hind wing looks like that number) flew a few feet above the lizards, and several turned their heads to watch it pass. Shortly after, another small butterfly of a different kind landed six inches or so from a lizard, which promptly lunged at it, made the catch, and gobbled it down.

Intrigued by the little drama taking place, Anne and I settled down to watch the action. Some of it was supplied by tourists throwing rocks at the lizards (even while we were obviously observing and photographing them). Despite these problems, in about two and a half hours of sunny weather, we saw hundreds of lizard reactions to butterflies, including about seventy-five attacks of which fifteen ended in a capture. On several occasions, lizards leaped clear off the ground to snatch up a passing butterfly. Although a few other insects were consumed, butterflies clearly were the main nutritional source for the lizards during our observations.

So what? Ideas about the role of the color patterns on the wings of Lepidoptera (butterflies and moths) have been important in the development of evolutionary theory. For more than a century biologists have realized that the patterns evolved largely in response to the selective action of visual predators on populations of Lepidoptera. Some species (such as the peppered moths) are assumed to have been selected to blend in with natural backgrounds so that predators would have a hard time seeing them. Other species have iridescent or otherwise very noticeable colors on the upper surface and camouflage on the lower. Such "flash and dazzle" patterns make the butterflies very prominent in flight. But by folding their wings and exposing only their camouflaged undersides when they land, they presumably fool predators by seeming to disappear in a fraction of a second. Some butterfly species display patterns, especially of orange or red and black, informing predators that they taste bad or are poisonous (these are often rejected by predators in the laboratory and, if eaten, may make them vomit). Still other butterfly species may mimic the wing-color patterns of the bad-tasting ones and thus often escape being attacked, even though they taste good themselves and are not poisonous.

You will recall that there is good observational evidence that natural selection for camouflage patterns is at work on the peppered moths. And laboratory experiments indicate the efficacy of mimicry. But until we got lucky at Iguaçu, there had been no direct systematic observations of predation on butterflies, and the attack patterns of predators on butterflies remained virtually unknown. Only a series of unconnected reports of individual attacks by birds and one of a lizard capturing a single butterfly had been published. Until then, evidence of predation on butterflies had been largely indirect; butterfly collections often contain specimens whose wings are clearly scarred with the outlines of bird beaks and lizard jaws, showing that these predators often attack butterflies.

Our observations were thus the first of a group of predators specializing on butterflies. The lizards were devouring butterflies at an impressive rate, fast enough to indicate that they were capable of placing considerable selective pressure on butterfly populations. Of the butterflies that were eaten by the lizards in the short time we watched, six were of the same *Callicore* species, and presumably from the same local population. Lizards and that *Callicore* species were both common around the falls, and the lizards probably were a major factor causing mortality in *Callicore* and other butterfly populations in that locale.

Significantly, the lizards did not pay equal attention to all butterflies: longwing butterflies, which are presumed to be distasteful and certainly are not inconspicuous, were ignored, while the *Callicores* (which belong to a group thought to be tasty) invariably attracted attention. Since this is what the theory of mimicry would predict, our observations helped confirm current ideas on the evolution of butterfly color patterns.

Sometimes the patterns of predator behavior that have evolved seem, from our viewpoint, to border on the bizarre. An example is found in some of the numerous species of cichlid fishes—relatives of the angelfishes so popular with aquarists—that live in Lake Malawi in Africa. You will recall them from our discussion of lek behavior in the last chapter. Many of these fishes are mouthbrooders. Males establish nests on the lake floor (often just a hollow scooped from the sand) and females join the males in the nest to spawn. After the eggs are laid, the male fertilizes them, and the female then promptly scoops them up in her mouth and swims away. She holds the eggs until they hatch, providing them with a constant stream of oxygen in fresh water taken into her mouth and expelled through her gills. And, after the eggs hatch, mama's mouth provides shelter for the small fry.

Analysis of the stomach contents of some other species of cichlids that are not mouthbrooders showed that they dine primarily on fish eggs and young. A series of observations by Ken McKaye explains how they get their meals. A predator carefully stalks a mouthbrooding female from either above or below. When in proper position, it lunges at the female and rams her in the head. This causes the mouthbrooder to spit out her young involuntarily, and the predator gobbles them up.

Several different cichlid species make their livings by head-ramming. They make up what ecologists call a guild—a group of different species that exploit a similar resource in a similar manner. You are already familiar with some other guilds, such as the guild of mammalian pursuit predators that run down hoofed animals, wearing out the prey in a long chase; wolves in North America and Eurasia and the hunting dogs on the African savanna are members of that guild. The guild of ambush predators, on the other hand, is made up of relatively short-winded animals that rush at their prey from cover; lions and leopards are classic examples. Other guilds are more obscure. Predacious aquatic insects have been divided into guilds on the basis of whether they seek out their prey or lie in wait for them, and whether they pierce their victims and suck out their body fluids or eat them whole.

Of course, prey have not remained evolutionarily inert in the face of the varied onslaught of predators. Instead, they have evolved various defenses—ways of reducing their chances of being eaten. The alertness, speed, and endurance of hoofed animals usually make their predators work hard for a meal, and as a result, those meals usually consist of the very young, the ill, or the decrepit. Alertness—ability to detect the presence of a predator—helps smaller prey as well. Cornell entomologist Barbara Peckarsky, working at the Rocky Mountain Biological Lab, has shown experimentally that prey of predacious aquatic insects can detect the presence of a predator chemically (both predators and prey are nymphs—wingless immature forms). In the ice-cold East River, which

The pattern on the underside of an alpine butterfly camouflages it on the lichen-covered rock on which it is sunning.

runs through the lab's land, she has placed transparent observation boxes which allow the water to flow through unimpeded. To determine the pattern of chemical dispersal within the box, Peckarsky first introduced a small screen tube containing food coloring and mapped the flow of the colored water. Then she substituted a tube containing predatory stonefly nymphs for the one with food coloring and mapped the distribution of prey insects such as mayfly nymphs that were introduced into the box. Prey species with the ability to detect the predator's chemical signs would move out of the "food coloring zone" when a stonefly was in the screen tube, and would move back in when the stonefly was taken out.

One of the most common forms of defense against predators is cryptic coloration—camouflage—as occurs in the peppered moths. It is certainly the most commonly evolved defense among smaller organisms such as insects, and is often found in large ones as well. A classic example of the latter is the spotted coat that makes an immobile fawn lying quietly on a sun-dappled forest floor almost invisible. Another is the pattern on the underside of an alpine *Oeneis* butterfly that makes it blend with the lichen-colored rocks on which it suns itself.

But sometimes the camouflage works not by blending the prey with its background but by making it look like something it is not. Some butterfly caterpillars have huge false "eyes" near the front end, making them look like small snakes—and thus a considerably less desirable morsel for, say, a bird. Some fruit flies have patterns on their wings that make them resemble a crab spider, dangerous enemies of many insect predators that might like to dine on the fly.

In other cases it isn't the prey but the predator that is cryptic or mimetic. One of the head-ramming cichlids is able to change color to resemble the prey species it is stalking, thus presumably throwing the prey off guard. Many spiders and some praying mantises sit in flowers that they match in color and snatch up visiting insects; one mantis "pretends" to be an entire flower and gobbles down inquisitive "pollinators." And, of course, mammalian ambush predators are often cryptically colored—crouching lions blend with the savanna grasses, and the stripes of the tiger break up its outline in the dappled sunlight of its forest haunts.

Cryptic coloration can, of course, help a moth to avoid being eaten in the daytime, but it is of no help against predators that hunt in the dark of night. The moths' most important nocturnal predators, bats, hunt insects by generating ultrasonic pulses—sounds with frequencies much higher than can be detected by the human ear. The returning echoes of the pulses indicate the location of flying prey, much in the manner that returning sound pulses from sonar equipment are used by surface vessels to locate submarines. Using these echoes in a process called echolocation, bats have an astonishing ability to assemble a picture of their environment—a picture so detailed as to allow them to identify the species of prey.

In response, some moths and other insects have evolved ears that detect the bat sonar so that they know when they are being hunted (just as subs have sonar receivers to detect *their* hunters). Such moths can take evasive action and have been shown to be 40 percent less likely to be caught than those without ears. Some moths go even further—they attempt to "jam" the bat's sonar with their own ultrasonic sounds.

The bats, in their turn, have evolved countermeasures to the moths' defenses, such as shifting their sonar frequencies out of the frequency range of maximum sensitivity of moth ears. Some bats even echolocate with frequencies entirely within the range of human hearing (where the moth ears are not sensitive), and a few have shifted their search tech-

Bombardier beetle *(Stenaptinus insignis)* spraying. The beetle is being secured by a wire glued to its back, and its leg is being grasped by a forceps to imitate the attack of an ant.

nique from echolocating to simply listening for sounds made by their prey.

A fascinating addendum to the bat and moth story is provided by moth ear mites. These tiny relatives of ticks are parasites of the moths, living in colonies in moth ears, and in doing so, destroying the ear's ability to hear. But, of course, a deaf moth is more likely to be eaten by a bat, which would mean the end of the ear-mite colony as well. The mites have therefore evolved behavior to assure that only one of a moth's two ears is infested, permitting the host moth to keep listening in the dark for hunting bats.

Much more active antipredator defenses than those of the moths have evolved in some other insects. When a bombardier beetle is assaulted by an ant, it whips its abdomen around and hoses the predator with a boiling hot, noxious spray. Elegant studies by Cornell University biologist Thomas Eisner have revealed the intricate mechanisms that allow the beetle to produce the required chemical reactions and control the scalding spray without injuring itself. Tom has also uncovered an astonishing array of chemical defenses in other invertebrates, and he

1

The insects strike back. The maggots of a large horsefly (1) lie submerged in mud and seize spadefoot toads (2), which they kill by feeding on their body fluids. The electron micrograph (3) of the head of a maggot shows the hooked fangs with which the toads are caught. The elongate depressions in the front of the fangs are thought to be the openings of ducts from venom glands.

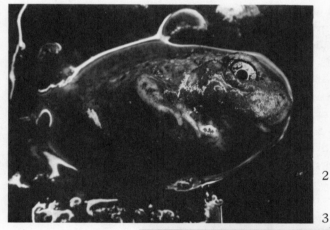

2

3

THREE PHOTOS:
T. EISNER, CORNELL UNIVERSITY

and his colleagues have documented the ultimate turning of the tables by an insect. Flies are standard items in the diets of many frogs. But the Eisner group discovered the other side of the coin: ferocious mud-dwelling maggots of a large horsefly that devour mature frogs!

In the face of such fascinating diversity in predator-prey interactions, the question of whether there are broad principles governing those interactions is of great scientific interest. For instance, one general question that has long fascinated ecologists is whether or not predators evolve prudence. A prudent predator would show restraint; it would not overeat, but rather would exploit its prey at a rate that gives the maximum sustainable harvest from the prey population. It is not clear, however, whether evolution can easily produce such behavior. Here again the question of altruism rears its head. In order for prudence to evolve by individual selection, individuals that showed restraint would have to reproduce more than those that did not. But obviously the wolves and lions that capture the most prey will have the most offspring and the best-fed offspring (which are then most likely to survive). Gluttony has its reproductive reward. Therefore individual selection would generally favor the individual predator that kills all the prey it can "convert" into offspring—regardless of the impact of its behavior, or that of its entire population, on the prey population.

If predators were prudent, then, one would have to look to group or kin selection to explain the evolution of their prudence. You will recall that both of these types of selection require such special circumstances that, at best, prudence would be expected to be a rare characteristic of predators. And rare it appears to be. Predators generally eat as rapidly as they can until they are surfeited. There are no documented cases of hungry predators not eating, even when the prey population is rapidly declining. Lynxes, for example, do not pass up opportunities to eat snowshoe hares when the hare populations are at low ebb.

Many factors in predator-prey interactions, however, tend to make predators "prudent" as a group, even though each individual is eating as much as it can lay its claws or jaws on. Prey individuals that are in prime reproductive condition, those that are most critical to the future of the prey population, are generally *least* likely to be caught. Also, in many cases, proportionately fewer individuals of a prey species are captured when its populations are very sparse than when they are very dense. This is because it pays a predator to specialize on a particular prey species only if it is abundant. Indeed, predators that frequently encounter one kind of prey (as they would if it were at high density) often form a search image—a preconceived notion of where to look for a common prey and what to look for—that helps them find more of the same kind of prey.

Ecologists doing fieldwork also often form search images. For instance, the caterpillars of many butterfly species are quite cryptic. Spotting the first few caterpillars in a field site is often difficult. But after a few are seen, suddenly the place may seem alive with them, although a few minutes earlier none were visible. The caterpillars haven't changed —the search image has been formed.

Birds apparently form search images too. Entire bird populations have been observed to start preferentially eating a certain kind of food when it became common and then switch to another kind once most of the first one has been consumed. There are also numerous reports of birds searching areas where they have previously found prey. Combined, the switching and searching behaviors define the formation of a search image in a natural population, since, of course, there is no way behaviorists can experience the perceptions of nonhuman organisms.

Bernd Heinrich, an ecologist at the University of Vermont who is famous for his important studies of foraging by bumblebees, has shown that birds can form search images for the leaf damage caused by feeding caterpillars. In an experimental cage, he and a colleague created an "artificial deciduous forest" using the top yard or two of white birch and chokecherry trees. They found that black-capped chickadees foraged preferentially in trees with ragged leaves that had either been partially eaten by caterpillars or had simulated caterpillar damage created by a paper punch. The chickadees were also able to distinguish the tree species. That discrimination is often important because some tree species are infested mostly by palatable caterpillars, while others play host mostly to noxious ones.

But the palatable species of caterpillars in Heinrich's study have not failed to respond evolutionarily to the birds' skills. Sometimes they chew the leaves cleanly, ribs and all, leaving no obvious tatters and shreds for the birds to spot, or cut off partially eaten leaves at the base or move away from damaged leaves. Unpalatable caterpillars, in contrast, usually eat only the soft part of the leaves, bypassing the tough ribs which then remain as telltale signs of their dining. A noxious caterpillar is presumed likely to survive discovery by a bird; a palatable one is not.

The formation and dissolution of search images thus can lead to proportionately less predator pressure on rare prey than on common prey. If predators concentrate on prey that is common, there is less chance of it driving a food source to extinction, which makes the system more "prudent." The formation of search images tends to keep predators and prey interacting with each other and influencing each other's evolution.

I commented earlier that it usually requires either luck or dedication to study predation systematically, but an exception to that rule is that special kind of predation known as "herbivory." It is no accident that predator-prey relationships in nature have been most thoroughly investigated in systems where the prey are rooted in the ground and cannot run away. In fact, some of the best-known fieldwork on animal-animal predation has been done in intertidal situations where the prey are barnacles, limpets, and other animals that are almost as immobile as plants. Where both predators and prey are mobile, the practical problems of studying predation in nature are formidable—just recall how difficult it is simply to determine the size of a *single* mobile population! As a result, ecologists have often leaned heavily on a combination of theory and laboratory experimentation to answer questions about interactions between mobile predators and prey. But these problems are avoided in studying plants.

I first got involved in investigating the interactions of plants and their predators long ago, as a sideline of my studies of checkerspot populations. In 1963, Peter Raven, now director of the Missouri Botanical Garden and an internationally renowned scientist, was a rising young plant evolutionist and my close colleague in Stanford's Department of Biological Sciences. One day we were drinking coffee at an ancient round table in the Dudley Herbarium, which housed the dried-plant collections of the university's Natural History Museum. This was, in fact, a daily morning ritual, with the cast of participants ever changing and the topic of conversation roaming from biology-department gossip to world events to ecology. It was the sort of forum in which a great deal of progress is made in science, possibly more than in the white-coated sterile laboratory situations so often depicted in the movies. On this particular day, I asked Peter a typical ecologist's question (in typical biologese). "Why the hell would *Euphydryas editha* feed on *Plantago* and *Euphydryas chalcedona* on *Diplacus*?"

My question is easily translated. I was curious as to why caterpillars of Edith's checkerspot butterflies fed on a plantain (a member of the plant family Plantaginaceae), while those of the closely related chalcedon checkerspot fed on a "scroph" (a member of the snapdragon family, Scrophulariaceae). "That's simple," Peter replied. "Don't you realize that the plantains are just wind-pollinated scrophs?" He was pointing out that the superficially different plants fed on by the closely related butterflies were in fact also closely related, despite their dissimilar appearance. Put another way, he was saying that the butterflies' dietary choices were as unsurprising as two brothers craving chocolate, but each preferring a different brand.

That little exchange led to an investigation of the relationships of butterflies to the plants they eat as caterpillars. It resulted in our co-authoring in 1964 an article, "Butterflies and plants: a study in coevolution," that played an important role in the development of one of the most active research areas in ecology today: the study of the reciprocal evolutionary relationships between ecologically intimate organisms such as predators and prey. The term we coined to describe those evolutionary relationships, coevolution, has become one of the buzzwords of contemporary ecology.

The basic theme of the butterfly-plant coevolution story actually dawned on Peter and me distressingly slowly. I described to him the patterns of food-plant use by caterpillars, and he looked for botanical patterns. Did most closely related butterflies feed on plants that were also closely related, or just similar in appearance, or growing in similar habitats, or what?

We found an abundance of raw data bearing on these questions. The eating habits of young butterflies were better known than those of any other large group of plant-eating animals. The reason is simple: butterflies are popular with collectors. Collectors like perfect specimens, ones with every colored scale on their wings still in place. The best way to get such specimens is to find caterpillars and feed them until they form chrysalises. When the adult emerges from the pupa and dries its wings, it is perfect. Its colors are glowing, and no scales have been lost to the vicissitudes of flying, seeking nectar, mating, and so on. The collector promptly adds the specimen to the collection.

But, in order to find and raise caterpillars, the collector must know what kinds of plants each feeds on. And butterflies are finicky eaters—most eat only one or a few species of plants. Thus collectors long ago started to amass information on the food plants of the caterpillars of all the butterfly species. As a result of this preoccupation of amateur lepidopterists, the dietary habits of most temperate-zone butterflies, and of a fair sampling of tropical ones, are well known.

By examining the literature on caterpillar diets and plant taxonomy, and integrating these with earlier work by others on plants and herbivores, Peter and I eventually saw an underlying pattern: the biochemistry of the plants, especially the diverse kinds of "secondary compounds" they contained, largely determined who ate what. We hypothesized that these complicated, energy-rich chemical compounds were not plant "waste products," as most botanists believed at the time, but actually chemical defenses evolved to ward off caterpillars and other predators of plants.

We concluded that patterns of food-plant use by caterpillars were largely created by the evolutionary success (or failure) of butterfly groups in overcoming the similar chemical defenses of related groups of plants. For example, a special set of nasty toxins are found in milkweeds, which makes them poisonous to most herbivores. But some herbivores, such as the closely related monarch and queen butterflies, have evolved resistance to those poisons and dine happily on milkweed plants. At some point, a common ancestor of the monarch and queen became capable of feeding on milkweeds, and now that shared capability is an indicator of their evolutionary relationship. Similarly, a set of toxins different from those in milkweeds characterizes cabbages and other members of the crucifer family, and those defenses have been penetrated by cabbage butterflies and their relatives. And apparently, longwing butterflies long ago evolved the capability of handling the kinds of defenses developed by passion vines, each species of longwing now generally attacking just one or a few species of the passion vines to which it is especially adapted.

In many cases, the chemicals in a group of plants that are distasteful or poisonous to most herbivores actually stimulate feeding by the caterpillars that have specialized in attacking that kind of plant. For example, cabbage butterflies have been found feeding on nasturtiums, plants unrelated to the cabbage which happen to have evolved the same kind of defensive chemicals.

Plants, of course, have special problems in defending themselves against parasites and predators. They can't swish their tails to shoo them away, avoid places where enemies occur, or run away if attacked. But they do have some nonchemical defenses. For instance, they can

What appear to be bits of leaf being carried along a branch by a column of leaf-cutter or "parasol" ants are actually the intact leaves of a small vine. Its appearance may be coincidental, but it is tempting to speculate that evolution has produced the resemblance because it helps protect the vine from herbivores. For instance, a herbivorous insect would not want to lay eggs on an ant's burden, since the bits of leaf are being carried to the ant's nest to serve as a growth medium for the fungus that the ants eat. The site is in the rain forest at the Organization of Tropical Study's field station, Finca La Selva, Costa Rica.

hide. In the deserts of South Africa are succulent plants that look exactly like rocks. And many plants in rain forests have leaves that look like those of other plants—presumably making it difficult for the herbivores that have learned to deal with their defenses to sort them out from look-alikes with different toxic chemicals.

Some plants have evolved even weirder deceptions. Once, in the rain forest at Finca La Selva in Costa Rica, I ducked under a branch that had a column of leaf cutter or "parasol" ants on it, each carrying a piece of leaf cut from a forest plant. These ants are the major herbivores of the forest, carrying huge quantities of leaf material to their giant underground nests, where the leaves are used as a medium on which the ants grow the fungi that they eat. As I passed this ant column, though, I noticed that it was not moving. On closer inspection, it turned out to be a vine—the shape, size, and spacing of its leaves closely simulating the leaf pieces carried by ants. It could just have been coincidental. But no herbivorous insect in its right mind would lay an egg on a piece of leaf being carried by an ant to a fungus farm. It is more likely that natural selection had given the vine some protection by shaping it to mimic a parasol-ant column.

Hiding or looking like something you are not are two nonchemical defenses open to plants. Another is the evolution of mechanical devices to make eating the plant more difficult. The spines of *Opuntia* (pricklypear) cacti are a good example. They certainly discourage most browsing animals, if not insect attackers.

But most plants, we now realize, have resorted to chemical warfare as their main defense against enemies. It is, in fact, the coevolution of plants with herbivores that is responsible for the enormous diversity of biochemical compounds found in plants. Those compounds are the sources of most of the mind-bending drugs used by people, the active ingredients of marijuana and cocaine being prominent examples. But people are not the only animals affected by psychoactive chemicals, which may have evolved specifically to alter the behavior of mammalian herbivores. After all, if a deer nibbles a hallucinogenic plant and then happily trots off into the arms of a cougar, it is unlikely to return to pester the plant.

In addition to the hallucinogens, plant defensive chemicals include acute poisons such as cyanide (found in some legumes) and nicotine (in tobacco of course), gums and resins (such as latex, which is the source of rubber) in which attacking insects can get trapped, and various chemicals such as tannins (in many plants, including oak trees and tea) that make the plants less digestible.

A classic study of the evolution of plant defenses, indicating the importance of toxins, was published by Daniel Janzen shortly after Peter's and my paper appeared. Dan, now the dean of tropical field biologists, investigated the relationship between certain ants and a group of Central American acacia trees that bear swollen, hollow thorns. The ants live in the thorns and feed on special tissues that the trees produce for them. The ants, in return, drive off herbivores and prune vines that attempt to climb the trees.

When the ants were removed from the trees, Janzen found that the trees were quickly overwhelmed by enemies. Trying a few leaves himself, Dan discovered that they tasted rather bland. In contrast, acacias not associated with ants had bitter leaves. Clearly the ant-hosting acacias, faced with defending themselves either with ants or toxins, found the expense of feeding the feisty insects lower than the cost of manufacturing the chemical defenses characteristic of other acacias. Interestingly, a small group of herbivores appears to have evolved the ability to deal with the chemical defenses of the antless acacias and specialize in feeding on them; another small group, including some beetles with armor too thick for the ants' jaws to pierce, has penetrated the ant defense.

Of course, all successful herbivores have found ways around the defenses of plants they feed on. The attackers can evolve resistance to the plant poisons in just the same way as our malarial mosquitoes or laboratory cultures of *Drosophila* can become resistant to DDT. One can imagine plants and the herbivores that attack them as being in a "coevolutionary race"—the plants being selected for better and better defenses, and the herbivores for the ability to overcome those defenses. Of course, it rarely is a simple one-on-one race, because some plants are attacked by many herbivores and many herbivores eat more than one kind of plant. A plant under multiple assault develops a compromise strategy that gives the best overall protection. Such many-on-many interactions lead to what is technically called diffuse coevolution.

One strategy against coevolving herbivores appears to be for plant populations to consist of biochemically diverse individuals, making it more difficult for herbivores feeding on that population to evolve resistance. This strategy was first recognized in the course of research begun in the late 1960s in a lovely subalpine meadow just below the Rocky Mountain Biological Lab in Colorado. In those days, the basic idea of plant-herbivore coevolution was still controversial. Some biologists argued that herbivores generally did not put enough selection pressure on plants to affect their evolution significantly. Thus Dennis Breedlove, a former student of Peter Raven's, and I decided to measure the impact of herbivores on plants in a natural situation.

For our study, we used a tiny blue butterfly (*Glaucopsyche lygdamus,* a member of a tribe of butterflies known as "the blues") which lays its eggs on unopened lupine flower buds. After hatching, caterpillars feed on the lupine buds and flowers. The eggs of this species of butterfly on an individual flower stalk could be easily identified and counted with the help of a hand magnifying lens. Two groups of fifty flower stalks were tagged and numbered, and the eggs on each were counted every few days (too often for an egg to be laid and the caterpillar to hatch unobserved). All the eggs were removed from one group of fifty stalks, thus providing a control group of stalks on which the flowers were subject to all hazards except being eaten by young blue butterflies. The remaining experimental group was subject to the attack of the butterfly larvae as well.

When the flowers had set fruit, we collected the stalks. For each of the two lupine groups, we determined the percentage of buds that had developed into flowers and that, in turn, had survived to produce fruit and seeds (even flowers that had been demolished left scars on the stalks that could be seen under a microscope). In the control (no blue larvae) group, 72 percent of the buds developed into flowers that produced seeds; in the experimental group, only 28 percent did. The interpretation of the experiment was clear: one small herbivore had destroyed nearly half of the reproductive potential of the lupine population, leaving no doubt that the blue butterfly could exert strong selective pressure on that population.

Although that was a significant piece of information, later work revealed something even more important: the defenses of different lupine species and of different populations within the same species were not equally efficacious. For instance, populations of a different species of lupine suffered much heavier damage than that in our original study, with more than 85 percent of the flowers destroyed by the caterpillars, while other populations suffered less.

What could explain the differences? Our first hypothesis was that the lupine populations that suffered least from the attacks of the blue butterflies would have the highest levels of defensive chemicals. Lupines were known to contain a variety of alkaloids, a group of chemicals that includes such poisons as nicotine, cocaine, caffeine, and strychnine. Analyzing alkaloid content of various lupines, therefore, seemed like a good place to start looking for a reason for their different susceptibilities to herbivores.

Careful analysis of the alkaloidal content of different lupine populations produced a surprising result, however. The primary difference between heavily and lightly attacked populations was not in the amount but in the *variety* of alkaloids present. Each plant in one heavily attacked population had a mixture of nine different alkaloids, and that mixture did not vary significantly from individual to individual. In contrast, each plant in a population that escaped heavy predation contained only three or four alkaloids, but the mixture of kinds of alkaloids in each individual was different. For instance, if one lupine plant had alkaloids ABCD, another might have DFG, another ABEH, and another CDH.

Presumably, the strategy of plant-to-plant biochemical variation in the lupines made it more difficult for the blue butterflies to evolve resistance to their poisons. A blue spends its entire larval life on a single plant. The surviving caterpillars from all the eggs laid on a given plant would be those most resistant to the particular combination of poisons in that plant. But when those survivors became adults, they would fly around and mate, and the females would lay their eggs on many different plants. Since most individual plants in the population are biochemically different, most of the larval offspring of those survivors would grow on plants with alkaloidal mixes different from those that their parents had been exposed to as caterpillars. Each generation thus is presented with different selective pressures (unlike the *Drosophila* exposed each generation to DDT), and evolution of the ability to deal with the poisons is thereby slowed.

Other work has since confirmed that the strategy of individual-to-individual biochemical variation can enhance a plant's defenses. This may be one reason why the gigantic populations of genetically relatively uniform plants created by human intervention—such as most crops—are relatively susceptible to the attacks of herbivores.

Plants evidently have been waging rather sophisticated chemical warfare against insects and plant diseases for hundreds of millions of years. Little wonder, then, that herbivorous insects usually adapt so easily to the crude attempts of *Homo sapiens* to poison them. Repeated exposure to the same insecticide (as I discovered in Bob Sokal's lab) ordinarily leads to the rapid evolution of resistance in herbivorous insects. Understanding the coevolutionary war between plants and insect herbivores puts the whole problem of predator-prey interactions and chemical pest control in a different light.

In contrast to herbivorous insects, carnivorous ones are usually hit relatively hard by pesticides. For one thing, they usually have much smaller populations (for reasons I will discuss later), and these are more likely to be exterminated. The reason for this is that if an insecticide kills an average 99 percent of all individuals, it would probably wipe out a population of one hundred predacious insects (because it would leave only one survivor), while it would leave about a thousand survivors from a population of ten thousand herbivores. So the herbivores would be able to restore their previous population size in a generation or so, while the predator might have to wait years for migrants to re-establish its population. The herbivores also benefit from the operation of Volterra's principle, which you will recall predicts that predators will be more damaged by insecticide spraying than prey because of the way the two populations interact.

But the herbivores also have a coevolutionary advantage when a field is soused with pesticides. They are already likely to possess adaptations for dealing with plant poisons, adaptations that can be modified for the poisons humanity invents. It is not surprising, then, that attempts to control insect pests with repeated spraying usually have undesirable results: since they are resistant to the poisons and now relatively free from the predators that normally help to control them, the pests swiftly rebound, often in greater numbers than were present before spraying.

But there is yet another undesirable result of such broadcast spray programs, one easily predicted from a rudimentary knowledge of predators and their prey: the appearance of brand-new pests. The decimation of predator populations means the destruction of natural controls, "releasing" previously unnoticed populations of herbivores. In a classic case of overspraying in cotton fields of the Cañete Valley of Peru, the spraying first worsened the problems with the old pests. Then it led to the "promotion" of an array of previously rare herbivores to pest status, as their normally small populations became gigantic following the poisoning of their predators. Another example of new pests created through the use of broadcast pesticides are the red spider mites. This group of free-living relatives of the inhabitants of moth's ears was promoted to pest status in many parts of the world by overzealous spraying of DDT, to which the mites' insect predators were more susceptible than were the mites. As a result, there is now a booming market for miticides where one did not exist just a few decades ago. Without doubt, the promotion of harmless insects to pest status has been a bonanza for manufacturers of insecticides. Their "cure" for resistance is more and more spraying—and thus a larger and larger market for their products, since each new pest that is created by overspraying becomes a target for new pesticides.

To be successful, control programs must be designed to avoid altering ecological relationships in undesirable ways. Pesticides *can* be extremely helpful tools when properly employed, but they must be used as scalpels, not bludgeons. So must nonchemical "biological" techniques of pest control, including predator enhancement programs. With these tools, ecologists are now able to design programs of integrated pest management that avoid the pitfalls of the dominant spray-spray-spray programs beloved of the pesticide industry and its subsidiary, the United States Department of Agriculture. How widely such programs will be adopted in the face of powerful economic interests remains to be seen.

A growing understanding of plant-herbivore coevolution, as you can see, has helped ecologists to understand a wide range of phenomena. It explains why plants harbor a multitude of biologically active chemicals—compounds people use as drugs, medicines, and spices. It also explains the origins of industrial products such as rubber. It has thrown light on such highly practical issues as why the promiscuous use of synthetic pesticides doesn't work in the long run and how best to control pests of crops with the least risk to human beings. It has helped to stimulate the investigation of other coevolutionary systems, such as those involving hosts and parasites, carnivores and prey, and groups of competitors. And it has advanced the complex task of unraveling how biological communities evolve.

Ecologists are now fully sensitized to the intricate development of adaptations and counteradaptations that appear to have shaped many coevolutionary systems, such as the bat-moth-mite complex described earlier. These systems now can be seen as coevolutionary races, in which extinction is the penalty for losing. In short, the idea of plant-herbivore coevolution did what a scientific theory ought to do: it provided general principles that explain diverse and previously unconnected observations. *All* of the kinds of interactions discussed in this chapter are now viewed by ecologists from the perspective of how they may have coevolved.

But that's not why I take a rather special pride in my collaboration with Peter on the butterfly-plant project. Rather the reason is that, in the course of the research, we two biologists only talked to each other and looked in the literature. Neither of us ever consulted an organism, living or dead, in museum, laboratory, or field. We had produced a counterexample to the old saying among biologists, "Study nature, not books." Our success in this case was based on studying books, not nature!

As is often the case, understanding one aspect of nature's machinery led to the illumination of another. Work that our research group was doing on butterfly-plant coevolution stimulated my thinking about a phenomenon Anne and I had observed in a habitat completely free of both butterflies and their food plants: the coral reefs of the Caribbean. It happened as we were taking a short vacation at Palm Island in the Grenadines on the way home after fieldwork in Trinidad. Naturally, we spent much of our time snorkeling. During the day, a resting school of fishes was a prominent feature of a small reef fifty yards from our cabin. It consisted mostly of young grunts, which spend the hours of darkness scouring sand flats and sea-grass beds for worms, crustaceans, mollusks, and other invertebrates. At dawn they returned to the reefs and remained there almost unmoving until they sallied forth again at dusk.

What attracted our attention was that the school was made up of more than one species of grunt and very often included individuals of yellow goatfish and, occasionally, snappers, members of entirely different fish families, as well. Furthermore, all of the fishes in the school were roughly the same size and had similar undramatic color patterns. The spotted goatfish, however, which is marked with three conspicuous dark blotches, was never found in the schools, even though it was common in the area.

Ordinarily, we might just have noticed the resting schools and thought no further about it. But coevolution was on our minds, and here were some ecologically intimate organisms of very different species that looked alike and hung around together. Why? Our observations suggested that the grunts and other fishes retreated to the reef during the day because its convoluted surface provided some shelter from open-water predators. We concluded that they formed schools as an additional protection against predators. The several species had coevolved similar color and behavior patterns, permitting them jointly to form much larger schools of similar fishes than would have been possible with the population of any one species alone. The uniform appearance of the individuals in the schools is important, as there is evidence that an "oddity" in a group elicits predator attack—and the individual that looks different tends to get eaten.

Some of the most interesting work on the coevolution of predator-prey systems in recent years has been done not on living predators and prey but on long dead ones. There has, in fact, been an upsurge of interest in the ecology of communities of the distant past. Fossil ecology is not only interesting for its own sake, but it can throw light on problems of general interest, such as the punctuation-gradualism controversy. For instance, paleontologist Robert Bakker has studied evolutionary patterns

Resting school of French grunts (majority, with diagonal stripes), smallmouth grunts (parallel stripes, at upper right), and goatfishes (single stripes, mostly near center) on a coral reef at St. Croix, U.S. Virgin Islands. The three species are able to make larger schools together than separately, and thus presumably gain more protection from predators. Note how they have evolved similar patterns so that no individual stands out from the group, and how only fishes of roughly similar size school together.

in 40-to-60-million-year-old fossil mammalian pursuit predators (the ecological analogs of today's wolves, hunting dogs, and hyenas) and their presumed prey. Careful analysis of leg bones from populations of different geological ages indicates that, while both pursuers and pursued evolved into more rapid runners, the prey sped up faster than the predators. This created what is called an adaptive gap. One species of predator apparently did not increase its speed at all for about half a million generations, allowing the gap to widen. Then that species disappeared from the fossil record and new pursuit predators replaced it. These then became more rapid runners and the gap began to close. Bakker called this kind of situation "the paradox of slow, stuttering coevolution." If the fossil record accurately reflects events of the past, then evolutionary ecologists must seek an explanation for why the increased speed of the prey did not place enough selection pressure on the first group of predators to let them "keep up" in the coevolutionary race. Might the habitat have been such that they actually were functioning more as ambush than pursuit predators?

How predators and prey run the coevolutionary race is of particular interest when the predators are so small that the prey is their entire environment—that is, when the predators are of that special kind we call parasites or pathogens and the prey are their hosts. It has generally been assumed that parasite-host systems coevolve so that the parasite does progressively less harm to the host. In other words, the parasite should become less virulent, and the host should evolve a better toler-ance for its presence. This seems logical, since otherwise the parasite is in danger of destroying its own environment.

Coevolution does often seem to go in this direction; in many cases, parasites attacking host populations that have long been exposed to them seem less lethal. In over four million square miles of middle Africa, newly introduced cattle quickly sicken and die. They are victims of nagana, a sleeping sickness of cattle caused by a trypanosome (a pro-tozoan that, like malaria, invades the blood). This protozoan is closely related to the agent that causes human sleeping sickness and, like the latter, is carried by tsetse flies. Varieties of cattle that have been in regions infested with nagana for a long time, such as the small, hump-less N'Dama cattle in West Africa, show some resistance. Native ante-lopes, on the other hand, which have been coevolving with the trypanosomes for millennia, harbor them with no apparent ill effects. The impact of parasite on host has clearly been softened by the long association.

Host-parasite coevolution does not necessarily follow such a path to accommodation, however, as Robert May of Princeton, one of today's leading ecological theoreticians, has pointed out. Virulence of the para-site is tied to how it is transmitted. Many virus parasites of insects, on one hand, are transmitted only when freed by the disintegration of the bodies of their hosts. In this case, evolution for decreased virulence would be fatal for the parasite. Nagana, on the other hand, is carried from one host to another by the bloodsucking tsetse flies, so there is no need for them to harm their hosts.

Human beings, of course, are hosts for many parasites, some of which cause serious diseases. Some, such as the viruses that cause influenza, measles, AIDS, and herpes, and the bacteria that produce scarlet fever, syphilis, and gonorrhea, are transmitted directly from per-son to person. Their ecology is relatively simple, and defenses against them involve preventing contact between carriers and noncarriers and either encouraging the natural defenses of the immune system (by immunization against viruses) or administering antibiotics (to fight against bacterial infections).

A number of other parasites that attack people spend part of their lives with nonhuman hosts. Four such species are the most important parasites of human beings: the protozoan blood parasites that cause malaria, all members of the genus *Plasmodium*. The other hosts of malarial parasites are mosquitoes of the genus *Anopheles*. Without both mosquito and human hosts, the parasites cannot survive.

Malarial organisms vary in their virulence, but generally do not kill their human hosts outright (as would be expected since death of the host is not necessary for transmission). Interestingly, it is the formation of the transmission stage in the blood that is the most debilitating phase of the infection. The parasites reproduce inside red blood cells and eventually burst them, releasing the infective stage (hopefully to be sucked up by a mosquito) and fever-causing waste products into the bloodstream.

The reasons for differences in virulence among malarial strains is not certain, but human populations heavily exposed to the most virulent of malarial organisms have evolved some natural defenses against them. The best known is a change in hemoglobin, the important protein that functions as an oxygen carrier in the red blood cells. In some parts of Africa, many people have two forms of this molecule—hemoglobin A and hemoglobin S (in shorthand, AS individuals), instead of just the normal A alone (AA). For reasons too complex to describe in detail here, AS individuals have considerable resistance against the most dangerous of the four malarial parasites.

Unfortunately, in the course of normal Mendelian heredity, some individuals with only the S hemoglobin (SS) are inevitably produced. Those SS individuals suffer a fatal sickle-cell anemia—so named because their red blood cells often have an unusual sickle shape (that is also why the hemoglobin is designated S). The AS individuals are selectively favored over *both* of the other types—over AA individuals who lack malaria resistance and over SS individuals who suffer the anemia and die before reproducing. But because of the underlying genetic mechanism, all three kinds of individuals continue being produced in the population, with the SS type being the rarest. Thus a few individuals pay the price of the deadly anemia for the benefit of malaria resistance accruing to others.

Of course, besides biologically evolved defenses against malaria, human beings have mounted cultural defenses. One of the first was drinking an infusion of bark of the cinchona tree (a member of the coffee family) as a cure for tropical fevers, long before the causative agent of malaria was known. More recently the active ingredient of the cinchona bark, quinine, was isolated and used, and still more recently a variety of synthetic agents (e.g., chloroquine) similar in chemical structure to quinine have been created. The mode of action of these drugs is uncertain, but they seem to act as a *Plasmodium*-specific poison, somehow interfering with the metabolism of the parasites in the bloodstream. Here it seems, at least at first glance, that cultural evolution has outdone biological—since the price of malaria resistance in this case does not automatically impose death on others as in the sickle-cell system. But this may not be entirely so in the long run, as we shall see.

After the discovery of the life cycle of *Plasmodium* about a century ago, a major focus in malaria control efforts became ecological. Rather than just trying to cure sick patients with quinine, attempts were made to interrupt the parasite's cycle in two ways. One was by using screens on doors and windows and bed netting to prevent the mosquitoes from having contact with human beings. The second way was by altering the environment of populations of anopheline mosquitoes. Since the eggs, larvae, and pupae of the mosquitoes live in water, control efforts often involved draining swamps, eliminating standing water in containers around houses, and so on. Chemical defenses were used as well. These included spraying with various insecticides and applying oil to water surfaces in breeding areas. The insecticides poison larvae, pupae, and adults. The oil suffocates the larvae and pupae by blocking the breathing tubes that they poke through the water surface.

Understanding the ecology of various *Anopheles* malaria vectors (transmitters) has been absolutely essential to malaria control—and lack of attention to it has sometimes led to disaster. Robert Desowitz, a professor of tropical medicine, describes in his fine book *New Guinea Tapeworms and Jewish Grandmothers* (New York: Norton, 1981) how a switch from subsistence agriculture to cash-cropping of rice brought a malaria epidemic to the Demerara River estuary of Guyana. The flooded rice fields provided increased breeding area for the local malaria vector, and mechanization reduced the numbers of domestic animals, which were its preferred source of blood meals. Large numbers of hungry mosquitoes thus were forced to turn to *Homo sapiens* for sustenance, and the result was a great increase in malarial infections.

Malaria today is resurgent all over the globe. For one thing, mosquito populations have evolved resistance to pesticides—as a result both of chemical assaults directed at them and as a side effect of the widespread spraying against agricultural pests in general. And the *Plasmodium* also have become increasingly resistant to chloroquine and other compounds which people take internally in an attempt to poison them. In both cases, the basic evolutionary process is similar to the one by which Bob Sokal's fruit flies became resistant to DDT. Biological evolution seems to be catching up with cultural evolution, and unless humanity is very clever and concentrates on control measures that avoid resistance problems, it may find itself in very deep trouble.

Not all parasites are as difficult to control as *Plasmodium*. Some parasitic cycles can be broken with relative ease. That a little ecological knowledge is sometimes all that is needed was demonstrated by the first successful control program ever directed at a specific human parasite. In 1869, the Russian naturalist A. P. Fedschenko "conquered" one of the most annoying of human parasites, the Guinea worm. This parasite, known as the "fiery dragon" in the Bible, goes under the scientific appellation of *Dracunculus medinensis.* It is common in tropical Africa, India, and the Near East. The adult female worm, which may grow as long as four feet, forms a sinuous tunnel beneath the skin of her host. An ulcer forms at the end of the tunnel. When the surface of the ulcer is wetted, the worm extrudes its hind end and deposits live-born larvae in the water. For millennia the "cure" was to wet the ulcer, catch the worm's end in a split stick, and *very* gradually reel out the resisting parasite over a period of days. Great care had to be taken to avoid breaking the worm, as an often fatal bacterial infection could follow an unsuccessful extraction.

Fedschenko worked out the life cycle of the worm. When an infected human being wades into water, the worm discharges larvae from the ulcer. The larvae in turn are eaten by *Cyclops* or other tiny crustaceans that serve as intermediate hosts for the worms. After the worms have developed inside the crustacean, they become capable of infecting *Homo sapiens* when the crustaceans are swallowed with drinking water.

This cycle shows clearly why step wells provide a perfect habitat for transmission. With step wells, people wade into the water to fill their buckets, rather than lowering and pulling them up on ropes. Fedschenko realized that if step wells were walled up so that a rope became necessary, there would be no way for someone with an ulcer to enter the water and allow the parasite's life cycle to be completed. Thus in Egypt, he encouraged the walling up of step wells, which greatly reduced the incidence of the disease. Guinea-worm infestation remains quite common, though, in India, where step wells have religious significance and have thus been more difficult to eliminate. But faith need not overwhelm ecology. Both the introduction of crustacean-eating fishes and chemical treatments to kill the *Cyclops* can help make even step wells safe.

Not all parasites are tiny creatures that live in bloodstreams of vertebrates, burrow under their skin, or, like a common mite, infest hair follicles in the skin of most human beings. A parasite may be bigger than its host, as often happens with avian brood parasites, such as cuckoos. Cuckoos lay their eggs in the nests of other birds, and very often the cuckoo egg closely mimics the coloration of the host's egg. When the young cuckoo hatches, it unceremoniously shoves its smaller, weaker, nest mates over the edge, leaving the foster parents with only the cuckoo chick to feed. Frequently in a matter of days, the youngster becomes larger than its foster parents, which are too bird-brained to recognize the deception. The result is one of the truly tragicomic scenes in nature: a pair of foster parents desperately trying to keep up with the voracious appetite of a young cuckoo.

Like many other aspects of the coevolution of predators and prey, there are numerous unanswered questions about the relationships between brood parasitic birds and their hosts. For example, cuckoo populations usually parasitize several different host species, but each female cuckoo tends to be a specialist—parasitizing only one species of host and laying eggs that closely mimic those of that particular species. That this is advantageous to the cuckoos is clear. Foster parents recognize the deception and desert the nest much more readily when an egg is laid in the nest of a "wrong" host than when it is laid in one of a host whose eggs are closely mimicked. But the genetic mechanism that allows the coexistence in the same cuckoo population of females laying strikingly different kinds of eggs and parasitizing different hosts remains a mystery.

Also little understood is the impact of brood parasites on the dynamics of the host species. It would appear sometimes to be considerable, but detailed studies have rarely been carried out. Indeed, little is known about the precise role of most kinds of parasites in the dynamics of their host populations, although there is abundant evidence that their impact can be enormous. A widespread epidemic of rinderpest virus in

East Africa at the end of the last century seems to have had so large an impact that its effects can still be seen in the distribution of antelopes. And the literature on the biological control of pests contains numerous accounts of the introduction of parasites leading to enormous reductions in the population size of harmful insects.

The story of the black death (bubonic plague) attests that a parasite can dramatically change the sizes of populations of our own species as well. Between 1348 and 1350, the plague killed an estimated 25 percent of the inhabitants of Europe. In the second half of the fourteenth century, many European cities lost half or more of their inhabitants, and the population of England was reduced by some 45 percent. In fact, the current potentially disastrous population explosion of *Homo sapiens* can be very largely traced to the control of the parasites of humanity through public-health measures and the use of antibiotics—human death rates have been lowered without a compensating reduction in birth rates.

Parasites that directly attack human beings are not the only predators to have exerted a very powerful influence on both human population size and growth and thereby on the course of history. So have some herbivores. In the best-documented case, the herbivore was a fungus that thrives on a diet of potatoes, and the human population affected lived in Ireland.

By the 1840s, Ireland's population had nearly outstripped its agricultural capacity, and British occupation and exploitation had reduced the Irish to abject poverty. In one town in County Donegal, there were only 243 stools, 93 chairs, and 10 beds among 9,000 people. The land available to peasants was divided into tiny parcels, and their staple diet had become potatoes, which could produce six times as much food per acre as grains. "Elegant" foods like milk and butter were eaten only rarely, on festive occasions.

In 1845 a fine spring was followed by a very wet and cold summer, providing an ideal environment for the fungal parasite of potatoes, *Phytophthora infestans*. As a result, in that year about half of the Irish potato crop was ruined. Due to a lack of seed potatoes, only about two-thirds of a normal crop was planted in 1846, and *Phytophthora* destroyed nearly all of it. In 1847 there were even fewer seed potatoes, and only 10 to 20 percent of the usual acreage could be planted, so little food was produced. In 1848 essentially the total crop was again lost.

The events caused by these failures are known as the Irish Potato Famine. Between 1846 and 1851, out of some eight million Irish, about a million and a half people died of hunger and diseases related to hunger, and another million or so emigrated. British attempts at relief were criminally incompetent, and accounts of the suffering of the people are heartrending to read. The Irish population continued to decline after the famine, largely through emigration, and remains well below its 1840 size today.

Enemies of plants can also work to the great benefit of human populations, however. Prickly-pear cacti *(Opuntia)* were imported to Australia from South America by early settlers who apparently wanted to use them as ornamental plants. Unfortunately, the cacti found the environment much to their liking and spread over a vast area. By 1925, 60 million acres of Queensland and New South Wales were covered with cacti, half of these acres so heavily infested that the land was useless. The cost of removing the cacti mechanically or poisoning them with chemicals was more than the land itself was worth, so entomologists ransacked the American homeland of *Opuntia* to find an insect that might help control the pest.

The larvae of one small moth with a big name, *Cactoblastis cactorum,* proved to be voracious eaters of *Opuntia*. Within five years after being released in Australia in 1926, the moth had done a spectacular job of destroying the vast majority of the cacti. Square mile after square mile of thick cactus stands melted away before hordes of *Cactoblastis* caterpillars—a classic example of the successful "biological control" of a pest. Since that time, *Opuntia* has never again been a problem in Australia. It exists there only in small clumps that are eventually found by dispersing moths and destroyed.

In addition to being a spectacular case of biological (rather than chemical) control of a pest, the *Opuntia/Cactoblastis* story has considerable theoretical significance as another piece of evidence against the argument of some biologists that herbivores play only a minor role in controlling the size of plant populations. In this case, a small moth changed the size of the Australian *Opuntia* population from immense to minuscule and has kept it that way ever since.

A biologist unfamiliar with the history of the moth-cactus interaction would be hard pressed today to determine the reason for the rarity of *Opuntia* in Australia. Twenty years ago in Queensland, along with Australian ecologist Charles Birch, I searched in vain for *Cactoblastis* among scattered clumps of *Opuntia*. Once a clump is "discovered" by *Cactoblastis,* it apparently does not last long, so at any one time most of the cactus clumps that can be found are free of moth larvae. Consequently, a sampling would suggest that *Cactoblastis* rarely attacks *Opuntia,* and logically leads one to conclude that *Cactoblastis* has little to do with the distribution and abundance of the cactus—precisely the wrong conclusion.

As in the *Opuntia/Cactoblastis* case, one of the techniques of intelligent pest control is to enhance the population size or extend the distribution of a predator that eats the pest. Sometimes *Gambusia* fishes have been introduced into bodies of water to eat mosquito larvae. By dispersing the *Gambusia,* humanity enters into what ecologists call a "mutualistic relationship" with the fishes: we help them to invade new habitats and thus expand their population; they assault our enemies, reducing our annoyance and our rates of sickness and death. Both partners benefit. Although it is not as common as predation (all animals have to eat, after all), mutualism is widespread in nature.

Some of the closest mutualistic associations known are between plants and the animals that pollinate them. Each of some 900 species of fig plant, for instance, is pollinated exclusively by its own species of fig wasp. The flowers of the fig plants are tiny and grow *inside* the roughly spherical fig that people eat (which is not strictly a fruit, but closer to an inside-out bouquet). The tiny female wasp enters the fig, deposits pollen on the flowers, lays eggs in the flowers, and then dies. The wasp larvae grow inside the flowers, which they largely consume, and form pupae. The wingless male fig wasps emerge before the females and march around inside the fig looking for flowers containing females. They use their telescoping abdomens to copulate with the females before the latter emerge from their pupae. The males then die without ever leaving the fig. The winged females emerge, collect pollen from the remaining flowers, and leave that fig in search of another in the proper condition to begin the cycle again.

The figs (including the commercial varieties) and their wasps are utterly dependent upon one another. The fig sacrifices some of its tissue to the feeding wasps in return for being fertilized. The wasp cannot mature anywhere else. Thus if there were no figs, there would be no fig wasps, and vice versa. Indeed, the first Smyrna figs that were grown in California could not produce a crop of mature figs until the appropriate species of fig wasp was introduced.

The world of pollination is replete with stories of such fascinatingly coevolved mutualistic systems. For instance, most insects go to flowers to get food—pollen or nectar—and in the process pollinate the plants (which evolved attractive flowers and edible rewards to lure their pollinators). In tropical America, however, male orchid bees pollinate orchids in return not for nourishment but for chemical scents which they collect. The bees are thought to modify the flower fragrances into substances whose odor attracts female orchid bees. Other orchids mimic female wasps, and get pollinated when love-sick male wasps attempt to copulate with them.

The mutualistic connections between many plants and animals created by pollination systems are important in binding the machinery of nature into a functional whole. In addition to their role in the wild, however, pollination systems are critical to many kinds of agriculture. In the United States, about a hundred crops, from fruit trees to alfalfa, are insect pollinated. Without pollinators, humanity would be denied a substantial portion of its food.

Another important connection in nature's machinery is the mutualistic relationship between plants and the myriad animals that help to disperse their seeds. When turtles, birds, bats, tapirs, horses, chimps, or people eat fruits, they are often doing the plant a favor, because the seeds pass undamaged through the digestive tract and are deposited in a new location. But the plants return the favor by feeding the animal. Indeed, so important are some trees to the maintenance of entire guilds of fruit-eaters within ecosystems that they have been dubbed "keystone mutualists" by Larry Gilbert. And it is not only terrestrial animals that are mutualists with trees. It has recently been discovered in the Amazon Basin that fruit-eating fishes are important in dispersing the seeds of certain terrestrial plants. Indeed, the *reason* that plants produce edible fruits is to get animals to disperse the plants' offspring.

Sometimes the relationships of plants and their seed dispersers are not mutualistic, but more nearly parasitic, since the dispersers are not helped, and may even be harmed—as anyone knows who has tried to remove "sticktights" or burs from socks or pants. These fruits have evolved hooked processes for attaching to passing birds or mammals. The birds and mammals then carry the fruits to distant places, where they eventually drop off and the seeds within them germinate. Occasionally, animals, especially birds, will get so coated with clinging seeds that they actually perish as a result.

Once dispersed, the seeds themselves are subject to predation by a variety of seed-eating mammals, birds, and insects. It is therefore not

A brightly colored anemone fish shelters within the tentacles of its host anemone at Raiatea, in the Society Islands. Anemone fishes become immune to the stings of the anemones, and the anemones are sometimes fed by the fishes. Thus the relationship is mutualistic, since both sides benefit.

surprising that, in the course of evolution, some animals have been recruited to help bury seeds as well. One of the neatest tricks in the plant world has been the evolution by some species of special nutritious tissues on their seeds that attract ants. The ants drag the seeds into their nests, eat the special tissues, and leave the rest. Northwestern University professor Andrew Beattie, who is also the director of the Rocky Mountain Biological Lab, and a colleague at Northwestern, David Culver, have shown that seeds of these species that are transported to ant nests are three times as likely to germinate as those that are not, and the young plants growing on the nests are more than four times as likely to live for at least two years.

Other examples of mutualism are found on coral reefs. Anyone who has snorkeled extensively on reefs in the South Pacific cannot help but have noticed the orange-and-white or black-and-white damselfishes that hang in the water column over large sea anemones, or nestle within their tentacles. These anemone fishes feed on plankton and are thought to help the anemone by occasionally bringing it bits of food. Whenever danger threatens, they swiftly retreat into the sanctuary of the anemone's stinging tentacles, to whose venom the anemone fishes are immune.

Furthermore, most of the fishes on the reef benefit from a naturally evolved biological control program. Various species of brightly colored small fishes, especially members of the wrasse family in the Pacific and the goby family in the Atlantic, get food by "cleaning" other fishes. The fishes that are cleaned, in turn, are benefited by the removal of crustacean parasites and possibly also by the removal of loose scales and trimming of wounds (whether the latter two are a significant benefit is not clear). The "paradise fish" (the one in which females turn into males) is iridescent blue with a black stripe from head to tail. Its close relative in Hawaii is even more striking, since it has added a bright-yellow head to the basic pattern. Interestingly, the 1½-inch-long gobies that clean in the Caribbean are similar in pattern to the Pacific paradise fish, and, when fishes from the Pacific are placed in aquaria with the gobies, those fishes instinctively recognize the gobies as cleaners.

The cleaners set up stations to service the fishes of the reef, advertising their availability with a characteristic dance. The prospective clients, including large predators such as lizard fishes and groupers, approach the cleaning station and become immobile in a special posture that makes them appear to be in a trance and signals their readiness to be cleaned. The cleaner then picks over the other fish. The cleanee will even open its mouth so its teeth can be cleaned, and also gape its gills so that parasites can be removed there. The cleaners enter these cavities, even those of the predators, with impunity.

An additional actor has coevolved with the cleaner-cleanee system: saber-toothed blennies, relatives of those hole-dwelling dash-and-slash *Plagiotremis* blennies of the Barrier Reef, which closely mimic the appearance and dance of the paradise fish. They too set up stations on the reef, and they too attract cleanees. The latter assume their trancelike cleaning posture, and the blenny swims up and takes a big bite, then hangs in the water casually chewing it. So rudely assaulted, the cleanee twitches, but does not attack the blenny. Apparently its resemblance to a real cleaner inhibits the larger fish. And that resemblance is quite good. Once at Bora-Bora, my friend John Holdren was holding a live paradise fish in his hand while I photographed it. It proceeded to take a small chunk out of John's finger—the "paradise fish" was actually a mimic blenny.

Mimicry, of course, is an extremely widespread phenomenon. It is especially well studied in my favorite group of nonhuman organisms, butterflies, where, you will recall, tasty species often resemble nasty-tasting or poisonous ones that fly with them. In these situations, the mimics are rather like parasites on the models, but mimicry can be mutualistic as well. It turns out that noxious species that occur together often resemble each other. The idea in the latter case is that the noxious species all gain protection by reducing the number of color patterns on butterfly wings that simpleminded predators must learn to avoid. Young predators learn by sampling possible prey, and the sampling process can be tough on a butterfly even if it is eventually spit out. If many distasteful butterfly species have evolved the same garb, the predators waste less time learning what color patterns to avoid, and each of those species loses fewer individuals while teaching them. All of the butterfly species benefit from the association.

Predators and prey and mutualists are not, however, the only classes of organisms that are ecologically intimate. There is more to interspecies ecology than eating, being eaten, and cooperating. Often there is competition. Competition occurs when two or more individuals use the same limited resource in a manner that reduces the supply for at least one of them. This usually means that they both simply consume a share of the *same* resource (as in the case of food), but sometimes a competitor uses interference to increase its share (as when a territorial hummingbird chases butterflies away from nectar-rich flowers). Competition holds great fascination for ecologists, both because of its potential importance in the machinery of nature and because competition plays such a prominent role in the economy of many human societies.

Competition has been studied in many different experimental systems. One of the most interesting involves flour beetles of the genus *Tribolium*. Work on these small insects is an excellent example of how ecologists use laboratory work to illuminate natural systems. These beetles once fed on natural accumulations of organic matter, such as are found in rodent nests. But they have evolved an association with *Homo sapiens* that goes back a long way. When the tomb of a pharaoh, dead since around 2500 B.C., was opened, an urn containing milled grain was found among the funerary relics. In the grain were the dried bodies of some *Tribolium*—so perhaps human grain stores, in which *Tribolium* are now major pests, can be considered a natural habitat of these beetles.

The popularity of *Tribolium* as a subject for experiments on population dynamics comes in large part from the ease with which they can be kept in the laboratory. A quarter-ounce of whole-wheat flour in a glass vial is a universe for these tiny insects. A pair of the beetles can be placed in the vial and the growth of the population in that microcosm observed. The procedure is simple; once a month the beetles are sieved out of the flour with a silk screen and the individuals in various stages —eggs, larvae, pupae, and adults—are counted, and then fresh flour is put in the vial and the beetles are returned to it. Many replicates of each experiment can be done with relative ease: the food supply is standardized, the shape of each environment is an identical cylinder, and all the vials can be stored in a closet-sized chamber with constant temperature and humidity.

If, instead of a pair of a single species, pairs of each of two different flour beetle species, *T. confusum* and *T. castaneum,* are added to the vials and the same procedure is followed, a series of competition experiments can be conducted in identical microcosms. That is exactly what the late Thomas Park, a famous ecologist at the University of Chicago, and his students did for many years around the middle of the century. What they found was that one species or the other always "won"; eventually the victorious species would be alone in the culture. Park found that under certain conditions of temperature and humidity, specific combinations of hot, cool, moist, and arid conditions, the outcome of the competition was always the same. For example, if the vials were kept in warm-moist chambers, *castaneum* was always the victor; in cool-arid environments, *confusum* always won.

In other environments, however, the results were not so straightforward. In a cool-moist environment, for instance, *confusum* won in 71 percent of the vials, and *castaneum* in 29 percent. And, interestingly, the performance of each species when it was alone in these environments did not accurately presage how it would do in the competition. For instance, contrary to what one might have expected, in a cool-moist environment, *castaneum* alone was able to maintain larger populations than *confusum* alone.

The indeterminacy of the outcome in some environments clearly needed explaining. One hypothesis advanced by population geneticists was that there was genetic variability for competitive ability in both species. That is, some individuals had genotypes that made them better competitors than others. In the hot-arid environments, presumably most *confusum* could outcompete *castaneum,* but some *castaneum* were genetically better competitors than *confusum.* The outcome in any particular vial would thus be determined by which particular individuals (good or bad competitors) of each species happened to be the ones used to start the experiment. Further experiments demonstrated that genetics could indeed influence the outcome. Strains could be selected that had greater or lesser competitive ability (showing that genetic variability existed for that trait), and there were more consistent results when inbred beetles (which were less likely to differ from one another genetically) were used as founders.

That is not the entire story, however. More experiments in the mid-1970s by Park and his colleagues indicated that the outcome of competition in environments that produce indeterminacy can be influenced by chance fluctuations in population sizes during the course of the experiment. That is, one species or the other becomes more common simply by accident, and whichever species accidentally becomes more common has a competitive advantage. Overall the *Tribolium* work demonstrated to ecologists that not only could competitors displace one another, but that who displaced whom could depend on either the physical environment, the genetic composition of founder populations, or pure chance.

In addition to an important body of experimental work, there is also a useful mathematical theory of competition. This theory illuminates, among other things, the circumstances under which an additional species can join a group of other species exploiting the same resource in the same way.

While competition is easily demonstrated in the laboratory and easily explained by theory, no topic in ecology is more controversial than the degree to which competition (and coevolution driven by competition) helps to shape ecological communities—in determining who lives where and why, and who lives together and how. So we will return to the topic of competition in Chapter 6, where the composition of communities is considered.

As a final note for this chapter, however, I should point out that Park's work with competing Coleoptera (beetles) was not an end in itself. He, like most ecologists, wanted to know how the machinery of nature works and how the cog we call humanity fits into the mechanism. He looked for broader implications in the work he did and was one of the first to warn against the consequences of human population growth.

In a 1962 paper in *Science* entitled "Beetles, competition, and populations," he wrote about the population explosion:

> I am against it! I do not wish, however, to draw direct parallels between insects and man. But despite this reluctance several facts have emerged from the study of beetles in their flour which seem to have general currency. One of these is that overexploitation and intense "interference" are perilous and that the peril increases as the population increases.
>
> And there is another fact . . . The largest population, if exposed to stress, does not necessarily enjoy the best prospect of survival. Man, as we all know and pontificate, has the intellectual talent and the technical skill to avoid such coleopterous hazards. In short, he has the capacity to manage his own population and (of equal importance) to conserve those myriad other populations on which he depends. But one thing is certain. If man does not manage *his* biology, *it* will manage him.

CHAPTER 5

WHO LIVES WHERE, AND WHY

Biogeography

TODAY'S ORGANISMS are the product of a long and complex evolutionary-ecological history. So are today's physical habitats. It is not surprising, then, that to understand the distributions of modern plants and animals —their biogeography—one must often look to the past.

Dramatic differences between communities are sometimes due simply to historical accident. For example, the absence of penguins in the Arctic and of polar bears in the Antarctic apparently stems not from an inability to adapt to conditions at the opposite pole, but from the failure of each to reach the other polar region by crossing the hot tropics. Failure to disperse across barriers in both the past and the present is one of the major factors in explaining who lives where today.

Indeed, a significant part of the explanation for the distributions of today's plants and animals can be found in the physical history of Earth, which is also a history of barriers. Changes in the configuration of the planet's surface created and destroyed physical barriers and modified

171

the climate, and changing climate, in turn, created and destroyed physiological barriers. All these influenced where evolving organisms could move and survive.

That geographic features are ever changing is a relatively old idea. Almost since the first maps of the globe were drawn, the resemblance has been noted between the S-shaped curve of the eastern edge of the continents of the Western Hemisphere and the S-shaped curve of the western edge of the continents of the Eastern Hemisphere. As a consequence, there has been speculation for centuries that the two great land masses were once contiguous and have slowly drawn apart. There was even speculation long ago about how such continental shifts might occur. Benjamin Franklin proposed that Earth consisted of a dense fluid beneath a firm outer shell, and that the shell would be "capable of being broken and disordered by the violent movements of fluids on which it rested."

At the beginning of this century, German meteorologist Alfred Wegener proposed that the continents were not fixed in position, but moved around. He pointed out, among other things, that they consisted of lighter granitic rocks, while the ocean basins were made of heavier, basaltic rocks—so that the former might more or less float on the latter. He developed a complete scheme of continental movement, postulating a single ancient landmass that had broken up to form the present continents. But physicists of his day calculated that no known force could cause the granitic continents to plow like ships through the basaltic ocean basins. And so, until a couple of decades ago, Wegener's ideas of "continental drift" languished, and most geologists believed that the continents had been in the same position since Earth's crust had solidified.

In the 1960s, however, discoveries once again called into question the idea that the continents had been forever fixed in the same positions, and a new picture of an ever-changing planetary surface has since emerged. A wide variety of evidence now indicates that Earth is made up of several plastic layers beneath a 60- to 100-mile-thick crust. That crust is proportionately thinner than an eggshell and is fractured into a series of "plates" upon which rest the oceans and continents. Each of the plates moves about on the hot, elastic layers below. At the midocean ridges, new crust is constantly being formed from magma (molten rock) rising from below. Where two plates meet at deep ocean trenches near the shores of the Pacific and elsewhere, the edge of one plate is being shoved under the other, descending into the hot layers and melting back into magma. The whole process is called plate tectonics.

Wegener had been pretty close to the truth. The continents do drift around, but not *through* the crust—rather *along with it.* The pace is most leisurely, only an inch or so a year, and the process appears to be complex. For example, small chunks of continent from one plate often get attached to the margin of a continent on another plate. Just as Wegener thought, about 250 million years ago all of the present continents were jammed together into a single giant continent, which itself had been formed from previously separate continents some 150 million years earlier. That long-gone great land mass is now called Pangaea, the name originally proposed by Wegener. The breakup of Pangaea began some 200 million years ago in the Cretaceous period, when the earliest dinosaurs and mammals were appearing. Since then the continents have gradually moved to their present positions. In the next 200 million years, they will have drifted into a totally different configuration—and by then the mammals may have gone the way of the dinosaurs.

Knowledge of the distributions of living and fossil organisms contributed significantly to this revolution in geology that has since become of crucial importance to biogeography itself. Biogeographers had long noted peculiar distributions of related organisms such as that of the southern beech tree, *Nothofagus,* which grows in New Zealand, Australia, and the southern Andes. Another that was difficult to explain in the absence of continental motion was the occurrence of flightless birds in Africa (ostriches), New Zealand (kiwis and extinct moas), and Australia (emus and cassowaries). These distributions lent considerable weight to the notion that a single supercontinent must have existed in the Southern Hemisphere in the distant past.

Geologists now believe that such a huge continent, called Gondwanaland, was a giant fragment created when Pangaea first split in two (the northern supercontinent was called Laurasia). Gondwanaland itself started to come apart a little more than 100 million years ago, gradually isolating the members of many groups of organisms with "southern distributions," such as the southern beeches and flightless birds. This sequence also explains very nicely the otherwise strange distribution of the Triassic mammallike reptile *Lystrosaurus,* whose remains have been discovered in Antarctica, southern Africa, and India, but not elsewhere.

But how were these distributions explained prior to the middle of this century, when geologists were convinced that the continents stood still? Biogeographers speculated that the southern continents must once have been connected by land bridges to permit migrations that could have created those remarkable biogeographic patterns. Eventually, those land bridges were supposed to have submerged beneath the oceans. Such connections, crisscrossing vast expanses of the oceans and subsequently submerging, were hypothesized with abandon. Of course, it is now clear that the Southern Hemisphere land bridges were entirely imaginary.

Thanks to plate tectonics, the geographic configuration of oceans and land masses of Earth is gradually but perpetually being altered. Furthermore, the pattern of oceans and continents is a major determinant of climates, and, as the continents shift, climates also shift. Continental drift consequently has had dramatic effects on Earth's climatic machinery and therefore on all the environments of the planet. For example, Pangaea is believed to have been assembled some 400 million years ago near the South Pole and to have been largely covered by a gigantic continental ice sheet. The Northern Hemisphere, in contrast, was mainly covered by unbroken warm ocean—a habitat in which the rich marine life of the Paleozoic era, including animals such as jawless fishes, long-extinct trilobites, and strange shelled relatives of squid, thrived in water rich in oxygen produced by algae. Pangaea then inched its way northward, and its climate, and thus the climate of most of Earth's land mass, gradually ameliorated.

Another important climate change occurred about 110 million years ago, when a widespread surge of seabed spreading formed broad midocean ridges—submerged mountain ranges—which displaced enormous quantities of seawater, possibly raising the sea level as much as 1,500 feet. Shallow seas penetrated the continents, dividing Africa in two, making Europe an archipelago, and inundating the central plains of North America. In the United States today a similar rise in sea level would put New Orleans, Memphis, Dallas, St. Louis, Kansas City, Chicago, and Minneapolis under the ocean along with most of Florida and every coastal city.

This tremendous rise in sea level also caused a great increase in the surface area of the oceans, which spread benign marine climates over much of the planet. Large bodies of water change in temperature much more slowly than does the atmosphere—they remain cooler in the summer and warmer in the winter. Oceans thus tend to warm the atmosphere in the winter and cool it in the summer, making climates in land areas near their shores less extreme than those in the centers of the continents. The spread of seas over major portions of continents thus created favorable conditions for reptiles, which then came into their heyday. Herds of gigantic herbivorous dinosaurs walked the Earth, pursued by predacious dinosaurs, which were the most ferocious terrestrial carnivores yet produced by evolution. Giant pterosaurs soared from the cliffs bordering the shallow seas stretching over Texas, Oklahoma, and Kansas.

The spreading of the seabed subsided about 85 million years ago, and the shallow continental seas receded. The continents drifted farther toward the North Pole, and this, combined with new mountain ranges created by the previous surge of spreading—the Himalayas, Alps, Andes, and Rockies—chilled the continents, and increased the temperature differences between the equator and the poles and between the seasons. Whether changes in the sun's output or in the composition of the atmosphere also contributed to these climatic changes is not known. Such a dramatic, though very slow, change in global climate could have caused the disappearance of many groups of organisms 65 million years ago at the end of the Cretaceous period, as well as the general decline of the reptiles at that time. Although it now appears that this gradual climatic change was not entirely responsible for the extinction of the dinosaurs, many paleontologists feel it at least helped them on their way —as we'll see a little further on.

Our planet is clearly not a static ecological arena at all, but one perturbed by vast geological events. Populations have changed in size and evolved not just in response to the kinds of short-term variations experienced between seasons, years, decades, centuries, or even millennia, but in response to enormously greater changes in climate occurring on a time scale of tens or hundreds of millions of years.

Not only have the physical features of Earth shaped the course of evolution, but evolution has shaped the Earth's physical features. As a major example, recall that the presence of oxygen in the atmosphere today is a consequence of the evolution of photosynthetic organisms. Atmospheric oxygen, in turn, has dramatically influenced the character and form of the biosphere—the relatively thin layer near Earth's surface in which organisms live. One important effect of the creation of atmospheric oxygen was to make possible the production of a thin layer of ozone (a form of oxygen) in the upper atmosphere which blocks out dangerous ultraviolet light from the sun. Until the existence of an "ozone shield" in the atmosphere, organisms could live only beneath the surface of the sea, whose waters protected them from the ultraviolet rays. Indeed, for the first four billion years of Earth's history, organisms were entirely aquatic. Nothing lived on the land. The first terrestrial plants did not evolve until the Silurian period, some 430 million years ago, when Earth was already 90 percent as old as it is today.

The creation of the ozone shield allowed plants to invade the land and started a vast expansion of the biosphere. It paved the way for the building of habitats that today we take for granted—it is, after all, plants that give structure to terrestrial communities (as is implicit in terms like forest, savanna, scrub, and grassland, which we use to describe them). The formation of plant communities on land, in turn, changed the amount of sunlight reflected from those surfaces and modified the hydrologic cycle—movements of water between oceans, atmosphere and land. In so doing, land plants have altered Earth's climate. They have also changed the physical surface of land by helping to form soil.

And the plant communities themselves have changed dramatically through time. It was only 35 million years or so ago, in the Tertiary, that they began to take on their modern form. By that time, flowering plants, the most prominent feature of the vast majority of present-day habitats, had developed their modern diversity; mammals had more or less fully replaced the dinosaurs; and our own ancestors had reached the status of "ape."

Although the major patterns of change are largely agreed upon, there is considerable controversy today over the pace at which evolution has transformed these communities. Recall that the gradualist school of thought holds that evolution has occurred more or less continuously, although at varying rates. The evolution of higher taxa (the upper level of the taxonomic hierarchy into which organisms are classified—genera, families, orders, classes, and so on) is viewed by gradualists as simply the summation of millions of years of microevolution.

Recently, you will recall, the punctuationists have challenged this view. They claim that the gaps in the fossil record are too frequent to be explained simply as the result of accidents of sampling (failure of fossils to be formed or found). Their interpretation of that record is that it is primarily one of the "sudden" (which in geological time may mean "in a few hundred thousand years") appearance of new species or higher categories, followed by a period of stasis in which little or no evolution occurs, followed in turn by the rapid disappearance of many kinds of organisms in an apparent episode of extinctions.

Much of the paleontological evidence, however, does not fit the model of punctuated equilibria. For instance, the fossil record of *Homo sapiens* shows a very gradual transition from short, small-brained, upright creatures, typified by the famous "Lucy," through a sequence of taller, larger-brained, upright *Homo habilis* and then *Homo erectus* (Java and Peking people) to the tallest, largest-brained *Homo sapiens*.

Furthermore, considerable evidence indicates that speciation in many kinds of organisms is going on at the present time, as in the checkerspot butterflies. The present, remember, could just be an episode of unusually rapid speciation. Or, perhaps species now being formed may have low survivorship—especially if they deviate far from ancestral forms. The pruning action of species selection—the differential survival of species—might eventually produce a macroevolutionary picture of relative stasis if and when future paleontologists examine the fossil remains of the Automotive Age. On the other hand, just as a great deal of past speciation undoubtedly cannot be detected in the fossil record, much of present speciation may be undetectable as well. It seems highly unlikely, for instance, that fossils of Edith's checkerspot could be distinguished from those of the closely related chalcedon checkerspots even though they are indisputably separate species.

All of these factors contribute to the current uncertainty about the degree to which the pace of evolution is stately or fitful. Yet it is an important question for ecologists, since the answer has implications both for the rates of environmental change and the coupling of those rates to the evolution of species and entire communities, both in the past and in the future.

The problems of reconstructing past environments as well as the relationships between now-extinct organisms and those environments are exemplified by attempts to understand the ecology of the dinosaurs. From before Darwin's time until rather recently, dinosaurs were viewed as gigantic versions of lizards. In fact, the word "dinosaur" means "terrible lizard." Gigantic, seventy-foot-long plant-eating dinosaurs of the genera *Apatosaurus* (better known by its former name, *Brontosaurus*) and *Diplodocus* were thought to have lived mostly wading in shallow water to help their legs support their thirty- to fifty-ton bulks. When they did move onto the land, they were pictured as lumbering slowly and stupidly over the landscape, dragging their ten-yard-long tails behind them. The giant meat-eaters that fed on them were also thought to be sluggish. Indeed, the picture created by some paleontologists of the famous and (presumably) ferocious *Tyrannosaurus* and its relatives was of creatures able only to lie in wait for their robotlike prey and then leap on them much as a sunning modern-day lizard would lunge for a passing insect.

This view of the dinosaurs traces to two great anatomists of the early-nineteenth century, one French and one British—Baron Georges Cuvier and Richard Owen. They viewed dinosaurs as simply enlarged versions of modern reptiles—cold-blooded organisms incapable of the sustained activity characteristic of birds and mammals, and none too bright to boot. The idea that dinosaurs were essentially enormous lizards was very compelling and enduring, so much so that some paleontologists claimed that giant herbivores such as *Apatosaurus* could not have stood erect but must have had their bodies suspended between bent, splayed-out legs like modern lizards and crocodilians. This belief was manifestly preposterous, since the skeletal anatomy of the legs showed clearly that they had the same straight, pillarlike design for supporting great weight as those of an elephant. Besides, the rib cage of an *Apatosaurus* was so large that the gigantic creature would have required a groove in the ground to move along if its legs and its body were joined in the same manner as those of a crocodile. That serious scientists would propose such a scheme in the face of all of the anatomical evidence indicates how entrenched the dinosaur-as-lizard notion was.

In retrospect (but *only* in retrospect), it is hard to understand why the lizard model persisted. Recent years have seen a much publicized and controversial reappraisal of the dinosaurs. We now know that the giant herbivores stood erect on four legs, and their giant predators stood erect on two, balancing themselves with their long tails. Clearly they did not spend most of their time flopped on their bellies, resting as contemporary crocodiles do. Some scientists, especially paleontologist Robert Bakker, maintain that if the dinosaurs did not spend most of their time lying around, then they must have had a fairly efficient metabolic system to permit the intense bursts of energy (and rapid recovery from oxygen deficits) that are characteristic of large, fast, highly maneuverable mammals today. This reduces to a problem in fossil physiological ecology: did some or all of the dinosaurs have such an efficient metabolic system? Attempting to answer that question involves reasoning from clues in the fossil record and from what is known of the metabolism of living organisms related to dinosaurs.

Today's cold-blooded lizards are thought to be inactive most of the time in part because they have comparatively inefficient respiratory and circulatory systems. Those two systems are unable to deliver the quantities of oxygen required for the high-powered metabolism found in warm-blooded animals. Furthermore, lizards (as well as snakes and crocodilians), if they are to perform at all, must use behavioral responses, especially precise orientation to the sun, to maintain an optimum temperature that may be relatively high compared to inanimate objects in their environment. For many lizards this optimum is about 86° F.

But the enormous bulk of many dinosaurs precludes the possibility that they used the sun to warm themselves to operating temperature each morning. It would take a few hours of basking in the sun to provide even a small rise in their body temperatures—although the exact times are a matter of some dispute, and would vary with the animal's size, shape, routing of the circulatory system, color, and so on. The basic reason is that the surface areas of large dinosaurs were too small relative to their volume. In addition, experiments with alligators, as well as theoretical calculations, indicate that in order to warm its viscera (so they can carry out their physiological functions) by basking in the sun, a large dinosaur would have to cook the tissues near the surface.

For these and other reasons, Bakker and some other scientists believe that dinosaurs were warm-blooded.* One further reason to think the dinosaurs were warm-blooded comes from attempts to reconstruct food chains that included them. Since modern cold-blooded reptiles do not require nearly as much energy per hour as mammals of the same size, reptilian predators can be much more abundant relative to their prey than mammalian predators. A population of rats can support more snakes, which need eat only one rat per month to survive, than it can weasels, which must eat a rat every few days. But in the late Cretaceous, for example, remains of the fossils of the herbivorous duck-billed hadrosaurs and rhinolike ceratopsians vastly outnumber the carnivorous dinosaurs that presumably fed on them. The ratio of the weight (biomass) of predators to that of prey appears more similar to that of lions and antelopes of the African veld than of a lizard population and its insect prey.

One must be cautious, however, in accepting this predator-to-prey-

---

* Note that, although I use the familiar term "warm-blooded," the more biologically correct distinction focuses on the source of the heat that the animal uses to reach operating temperature—in this case, warm-blooded would mean the heat was generated metabolically. "Cold-blooded" is used here for animals that reach operating temperature by taking in heat from their environments.

ratio argument for warm-bloodedness, because there are serious problems in treating the fossils from a geological stratum as a random sample of a community. Herbivorous dinosaurs may have spent more time in habitats where their remains were more likely to be fossilized than did the carnivores that preyed upon them. It is also possible that the largest herbivorous dinosaurs, like the elephants (which have evolved a similar life style), were relatively free of predation. If the dinosaurian carnivores had to dine mostly on the smaller herbivores, then the predator-prey ratio would have been much closer to that of lizards and their insect victims than the fossil record indicates. These possibilities have led other scientists to conclude that the food-chain evidence does not support the "warm-blooded dinosaurs" hypothesis.

Although paleontologists no longer believe that dinosaurs walked on splayed legs, the view that they were, in fact, mostly cold-blooded is still widely held. Some paleontologists argue that large size alone could produce a relatively high and constant body temperature in a mild, equable climate. And, they suggest, large dinosaurs could have remained in warm climates by seasonal migrations, while smaller ones could have become inactive if the weather became too hot or too cold. Furthermore, they claim that warm-bloodedness would have been too "costly" a trait, requiring more food than could be provided by the arid Triassic environments where dinosaurs thrived. Indeed, some believe that cold-bloodedness is a *preferred* trait in very large animals in tropical climates.

Underlying the whole controversy is the question of whether it is proper to extrapolate from the characteristics of modern lizards, crocodiles, and mammals to a variety of "imperatives" for ancient dinosaurs. For example, just because mammals are both highly successful and warm-blooded, does this mean that warm-bloodedness is necessarily always an advantage for them? Elephants might have been successful (at least until the rise of industrial *Homo sapiens*) *in spite* of being warm-blooded. And just because crocodiles are cold-blooded and sprawl while elephants are warm-blooded and erect, it does not automatically follow that the erect posture of *Apatosaurus* is evidence of warm-bloodedness. Nor is it required that all dinosaurs must have had the same patterns of heat generation and temperature control; some groups of fast-running dinosaurs may have been warm-blooded, and other dinosaurs may have been cold-blooded. Even modern mammals show variation in their degree of warm-bloodedness.

There have been several other interesting speculations about the evolutionary ecology of dinosaurs. Rather than sluggish, stupid creatures, dinosaurian predators—especially the smaller ones—may well have been swift, well-coordinated, relatively clever hunters. Recent studies using casts of the brain cavities of smaller predatory dinosaurs suggest a capability for rapid motor coordination and social behavior closer to that of an ostrich than a reptile. Such intelligent, well-coordinated carnivores could have placed considerable predation pressure on the small, nocturnal Mesozoic mammals. Indeed, some dinosaurs may well have specialized as nocturnal or crepuscular (dawn and dusk) hunters, snatching up our small ancestors and, for a period of time, limiting the evolutionary diversification of the mammals.

It also seems likely that in the rather equable climate of the Mesozoic, only the smallest dinosaurs (if they were warm-blooded) would have found it necessary to develop insulation to retain their body heat. The mere bulk of the larger species would have been sufficient to enable them to conserve heat through the cooler parts of the day. Some now think that the aerial cousins of the dinosaurs, the pterosaurs (pterodactyls and their relatives), were covered with long, silky insulating hair—and that feathers may have first evolved from this hair, also as an insulating mechanism, rather than an aid to flight.

If the dinosaurs were indeed warm-blooded (and we may never know for sure), they would be almost as different from the rest of the reptiles as are the mammals. Because of this, it has been suggested that the dinosaurs be placed in a separate class of vertebrates along with their descendants, the birds. Thus, instead of half-jokingly referring to birds as "feathered reptiles," as they have in the past, biologists could view them as "feathered dinosaurs"—the last survivors of an animal group that dominated Earth for more than 130 million years, successfully barring the mammals during that period from most of the ways of life they now pursue.

As you may suspect, the greatest ecological puzzle posed by the dinosaurs is: What caused the extinction of such a successful group of organisms? Some believe that they had been gradually declining before they disappeared some 65 million years ago, at the end of the Cretaceous (the last period of the Mesozoic era, an era that started some 225 million years ago and was followed by the Cenozoic era—the era we are still in today). Others contest this, claiming the evidence does not support a gradual decline. They point out that many dinosaurs disappear from the fossil record in a geological twinkling of an eye. And many other groups, such as the ammonites (mollusks with spiral, nautiluslike shells) and various marine phytoplankton, bowed out suddenly with them.

Few, if any, features of the geological record have been the subject of more speculation and controversy than the extinctions at the boundary between the Mesozoic and Cenozoic eras. Some of the speculation has been nonsensical, such as the notion that dinosaurs as a group became "senile" and died of old age, or that mammals wiped them out by eating their eggs (why, then, didn't they wipe out lizards, turtles, and crocodilians?). Other theories are related to the change in flora at the end of the Mesozoic. It has even been suggested that the decline of ferns changed the diets of herbivorous dinosaurs and that they died of constipation! More reasonably, the extinctions have been blamed on the failure of the herbivores to coevolve rapidly enough with the flowering plants, as the latter developed a vast array of antiherbivore chemicals. The carnivores that fed on the plant-eaters naturally followed them over the abyss.

None of the plant-related theories, however, explains why other groups (such as the phytoplankton) disappeared more or less simultaneously with the dinosaurs, why some groups of predacious dinosaurs could not survive by preying on mammals, or why at least some dinosaurs could not manage to keep up in the coevolutionary race with their food plants. Furthermore, since flowering plants appeared about 120 million years ago, why would their chemical warfare take 55 million years to conquer the dinosaurs and then do the deed with such catastrophic suddenness?

More tenable explanations for the sharp faunal and floral "punctuation" at the end of the Cretaceous involve the planetwide climatic change already mentioned. It has been suggested that in both marine and terrestrial environments, global coolings may have been the cause of a series of catastrophic extinction events in the past, including that at the end of the Cretaceous.

At that time, as we have seen, continental drift caused a cooling trend that could have had disastrous consequences for large, warm-blooded animals that, in the earlier, more equable climate, had not evolved insulating fur or feathers (just as today's elephants, rhinos, and hippos are essentially hairless). It would also have made things difficult for cold-blooded animals (amphibians and reptiles are scarce today in northern forests and absent from arctic tundras). Thus, whether they were warm- or cold-blooded, the dinosaurs would have felt the effects. The cooling could also have accounted for the marine extinctions. But the widespread nature of the extinctions argues against cooling being the sole answer. Surely some of the smaller, more rapidly reproducing dinosaurs could have evolved fast enough to avoid extinction if the cooling came on gradually, as it would have if continental drift were the overriding cause.

Various suggestions have been made for an extremely sudden event causing the extinctions, including a radiation "storm" created by a supernova in our portion of the galaxy, or a catastrophe caused by collision of Earth with an asteroid. The latter is the most popular recent theory. An asteroid perhaps six miles in diameter striking the planet would have killed many organisms outright with blast and heat and would have lofted huge amounts of dust into the atmosphere. The dust could have blocked the sunlight, causing very cold temperatures and turning off most or all of photosynthesis for months—ironically the same general sort of event as the nuclear winter that could follow a large-scale thermonuclear war. The survivors of such an event would have been primarily polar and temperate-zone plants and animals that had been dormant or in diapause or hibernation in the winter hemisphere, and plants that survived as seeds or underground structures elsewhere. Indeed, there is some evidence that such catastrophes, leading to mass extinctions, may occur at regular intervals—perhaps as frequently as about every 26 million years.

The asteroid hypothesis was proposed by a group of scientists headed by Nobel laureate physicist Luis Alvarez at the University of California, Berkeley, to explain some very thin and unusual geological strata (layers of sedimentary rock), which were deposited 65 million years ago. These strata, containing high concentrations of iridium, a metal similar to platinum, have been found in the Apennine Mountains near Gubbio, Italy, and elsewhere. They form a boundary between lower layers containing marine organisms typical of the Cretaceous and upper layers containing fossils from the Tertiary period. (The Tertiary is the first period of the Cenozoic era.) Normally, iridium is found only in trace amounts in the Earth's crust. It has a strong affinity for iron, and most terrestrial iridium presumably sank into the core of our planet when heavy and light elements were sorted out billions of years ago. Iridium, however, is relatively abundant in meteorites, which in combination with its abundance at the Mesozoic-Cenozoic boundary is what suggested the collision hypothesis.

In mid-1984, four scientists at the U.S. Geological Survey published a report showing that quartz grains in the iridium-rich boundary layer showed features characteristic of "shock metamorphism"—their crystal structures were the same as those found in quartz grains from the impact craters of meteorites, the sites of nuclear explosions, and laboratory shock experiments. The geophysicists concluded that their sample of quartz grains was "consistent with its origin as fallout from the impact of a large extraterrestrial body." But the question of the cause of the massive extinctions is still not settled. Other Earth scientists, who favor the view that a series of intense volcanic events deposited the iridium, point out that the quartz-grain evidence is consistent with their hypothesis as well.

If an extraterrestrial body did the job, where might it have come from? In 1981 a scientist at the Oak Ridge National Laboratory suggested that passage of another star close to our solar system could have triggered a bombardment of Earth with chunks of comets. Subsequently, other scientists have suggested that collisions with comets may be periodic, recurring about every 26 million years. The source of the periodicity was proposed to be an as-yet-undiscovered companion star of the sun that circles around it in an eccentric orbit, moving as far away as 12 trillion miles and coming as close as 3 trillion miles. When it is close, it passes through a collection of comets known as the Oort comet cloud, and its gravitation diverts some of the comets toward Earth, triggering a shower of comets that could last almost a million years. The hypothetical star, which has already been appropriately named "Nemesis," is not due to return for some 13 million years—which means that our civilization, at least, has more to fear from a nuclear winter.

Needless to say, this "periodic comet shower" hypothesis is also a subject of hot debate. It was proposed primarily to explain the conclusions of two University of Chicago paleontologists, David Raup and John Sepkoski. They did an extremely thorough statistical analysis of extinction events over the last 250 million years, using as basic data the appearance and disappearance of families of marine organisms in the fossil record. They carefully culled their data to remove groups of questionable taxonomic or temporal position, and compared the patterns revealed against two different estimates of the geological time scale. It was Raup and Sepkoski who found the most likely cycle (if there is a cycle) had a 26-million-year periodicity. Their hypothesis is further supported by statistical analyses of the ages of ancient craters formed by the impacts of extraterrestrial bodies. These appear to have a periodicity of about 28 million years—suspiciously close to that claimed for the extinctions.

Note, however, that the periodic-comet-shower hypothesis rests on a pyramid of assumptions involving the complex statistical analyses of collections of fossils and the proposed behavior of an as-yet-undiscovered star. Furthermore, it is well known that the type of analysis used on the fossil record can sometimes produce the appearance of regular cycles in data that actually have no periodicity at all. Therefore, at the moment, one must view the hypothesis that periodic mass extinctions are caused by periodic comet showers as a castle built on the foundations of an outhouse, but an interesting castle nonetheless.

Like the controversy over the cold- and warm-blooded dinosaurs, the debate over mass extinctions highlights the great fascination of reconstructing ecological relationships of the past, as well as the difficulties of doing so. But deciding whether catastrophic-extinction events have been relatively frequent in Earth's history is of great importance to ecologists and evolutionists. If they were, then chance physical events may have played a much greater role in the evolution of communities than has been previously suspected. Earth's entire biota would have been periodically confronted with bottlenecks, which only a rather small and perhaps random surviving sample of organisms managed to pass through. It could turn out that the reasons for the ascendance of the mammals actually had little to do with warm-bloodedness, the development of big brains, intensive parental care, or other adaptive "advantages." Instead it may have been simple luck—some of our ancestors survived the catastrophe at the end of the Cretaceous, and the dinosaurs did not. After all, mammals, with all their "advantages," coexisted with the dinosaurs for 140 million years, presumably providing sustenance for agile predatory dinosaurs. It was not until the dinosaurs were removed from the scene that mammals evolved their present diversity and took over the ecological roles previously played by the giant reptiles. So catastrophe for one group can be opportunity for another.

It is clear that, beyond catastrophes and continental drift, evolutionary history has been a major factor in determining who has lived where and when. That history includes the success or lack thereof of various products of evolution as a result of competition and dispersal from one area to another. We know, for instance, from the successful transplantation from one continent to another of organisms as diverse as English sparrows and *Opuntia* cactus, that many organisms can flourish in habitats that they had previously been unable to reach. Ecologists surmise that this is true for many, if not most, organisms. For example, penguins apparently evolved in the Southern Hemisphere, since no fossil penguins have been found outside the range of living species. As I mentioned earlier, they apparently have not managed to invade the Arctic due to the gigantic barrier of the warm, relatively food-poor waters of the tropical oceans. But there is every reason to believe that at least some penguin species could survive in the north polar regions if they could reach them.

Of course, we can't be certain of this. Perhaps a pioneering band of penguins did manage to migrate from the Galapagos (the northern limit of their present distribution) to Alaska. Maybe they made it at the height of a glacial period when the warm ocean barrier was somewhat narrower. And maybe for some reason the penguins just could not persist in arctic regions after they successfully reached them. There has been at least one casual transplant "experiment" in which nine king penguins were moved to northern Norway, and there were signs that they could have thrived there. But several of these transplanted birds, members of one of the largest and most brightly colored penguin species, were shot, and a population did not become established. Diving birds similar to penguins do thrive in the Arctic—indeed, the term "penguin" was first applied to a Northern Hemisphere flightless marine bird, the now-extinct great auk, a creature remarkably like medium-sized penguins. But no systematic attempt has been made to establish penguins in the Arctic.

This is probably just as well, since both accidental and purposeful transfers of organisms to places outside of their original range often have had unfortunate side effects. Goats, for example, have been moved to many parts of the planet far from their native haunts near the Mediterranean, and they can thrive in a great variety of communities. But often the communities into which they were introduced have been all but destroyed by the activities of these voracious herbivores. The island of St. Helena, in the South Atlantic, was once covered by heavy forests. Following timbering, however, introduced goats ate the seedlings that remained. This prevented regrowth of the trees, and converted the island into a rocky wasteland. (The ecological devastation of the Mediterra-

Penguins have never reached the north polar regions, even though they seem eminently suited to thrive there. Here a chinstrap penguin regurgitates food for its chick at Paradise Bay, on the Antarctic Peninsula.

Goats are perhaps the most destructive of the domestic animals humanity moves about the globe. Here they are contributing to the desertification of the Baringo region of Kenya. Note the large areas without vegetation.

nean basin itself is also in no small part due to the ravages of goats—in this case protected by people from natural predators in the area such as lions, which humanity exterminated there millennia ago.) Dogs, pigs, and rats, introduced when the Polynesians colonized the Hawaiian Islands between 400 and 600 A.D., helped eliminate a large portion of the native bird fauna long before Europeans arrived. And remember what *Opuntia* did in Australia. It is difficult to predict what the impact of moving penguins to the Arctic would be, but one possibility is the extinction (because of competition from penguins) of other species of sea birds that are less efficient predators.

Nevertheless transplant experiments remain a crucial tool for understanding biogeography, and it is possible to do safe ones, although on a less spectacular scale, to solve biogeographic dilemmas such as that posed by the penguins. Our group has been doing such an experiment with the rare Gillette's checkerspot butterfly *(Euphydryas gillettii)*, a relative of the Edith's and chalcedon checkerspots. Gillette's checkerspot occurs naturally in scattered colonies in the Rocky Mountains north of the Wyoming Basin (a low area splitting the Rockies in southern Wyoming from the Rockies in northwestern Colorado). We have been studying this checkerspot since the mid-1970s in the vicinity of the Grand Tetons in northwestern Wyoming.

Gillette's checkerspots are not found in Colorado, even though apparently suitable habitats occur near the Rocky Mountain Biological Laboratory and elsewhere. When we began our study, there were two possible explanations for the absence of this checkerspot species in Colorado. One was that the butterfly, having invaded North America from Eurasia (where its closest relatives live) via Alaska, was unable to cross the gap of unsuitable habitat in the Wyoming Basin and colonize the Colorado Rockies. The other was that it previously reached the southern Rockies, but the habitat there, although appearing appropriate, actually was incapable of supporting its populations.

My graduate student Cheryl Holdren and I carried out transplant experiments in an attempt to determine which of these two hypotheses is correct. If the butterflies *could* maintain populations in Colorado, that would be further evidence that checkerspots don't often use their flying ability to migrate long distances across unsuitable terrain (remember how little the Bay checkerspot moves around). On the other hand, if transplant attempts failed, then the absence of Gillette's checkerspot from Colorado would not necessarily indicate an inability to cross barriers. The matter was of more than casual interest because the ability of

Newly emerged female of Gillette's checkerspot on a leaf of twin-berry. This was the second individual seen at the Rocky Mountain Biological Lab, to which this Wyoming insect had been transplanted.

checkerspots to disperse may be a significant factor in whether or not enzyme variation in them is largely shaped by natural selection—since a very few migrants can greatly influence the frequencies of genes if they are "neutral" (see Appendix B).

In the summer of 1977, eggs and larvae were moved from Wyoming to Colorado. We placed them on their food plants, twinberry (a shrub in the honeysuckle family), close to the Rocky Mountain Biological Laboratory in an area that resembled their home habitat. The next summer, there was a small population of adults at the Colorado site, showing that the previous year's larvae had survived the winter in diapause, that the larvae had completed their development and successfully pupated in the Colorado spring, and that the transformation from larva to adult within the pupa had been successful. The behavior of the Colorado adults was similar to that in the source population, mating appeared normal, and egg masses were laid. But the butterflies in Colorado did not emerge until almost a month later than they did in Wyoming. By that time, so little of the growing season remained that we were afraid that many of their caterpillars would be unable to grow large enough to survive the winter, before the twinberries lost their leaves.

A visit to Colorado in early May of 1979 deepened our concern.

We had chosen our transplant site for convenience on a northeast-facing slope. In retrospect, this seemed clearly to have been an error. Snow had already melted off the sites where Colorado populations of Edith's checkerspot lived; they were mostly on south-facing slopes. While the larval food plants of those butterflies were already pushing above the ground, the little population of overwintering Gillette's checkerspot caterpillars and their twinberries were still buried under many feet of snow.

Cheryl and I therefore repeated our experiment that summer, but this time we transferred the butterflies to a site 1,000 feet lower and with a more southerly exposure (our choices were, of course, limited to places where the "proper" habitat occurred). In confirmation of our early spring fears, that summer *no* adult Gillette's checkerspots were seen at the original transplant site. Some evidently did fly, though, because we found a few egg masses.

In mid-May of 1980, we visited the Wyoming and low-altitude Colorado sites on consecutive days. Both had young caterpillars just starting to dine on leaf buds of twinberry. The two populations were perfectly synchronized. We were jubilant—success at last!

Those of you having any experience with fieldwork can guess how premature our elation was. Mother Nature always seems to have an ace up her sleeve for those who think they hold a better hand. Repeated searching late that summer turned up only a single female checkerspot and a single egg mass at the new low-altitude transplant site. In contrast, the seemingly doomed original population made a comeback—something on the order of two dozen adults were flying. In 1981 the lower site was, not unexpectedly, totally without Gillette's checkerspots. But the original north-facing, high-altitude location supported a population of more than a hundred individuals—and about the same numbers were present in the next four generations, 1982–1985.

So the experiment yielded an answer to our original question. A population was established that, after seven generations, was still persisting. The adult nectar and larval food resources appear adequate. No unrecognized predatory Colorado bug that devours Gillette's checkerspots has appeared. The climate permits survival. Thus it is most likely that the inability of the butterfly to cross the barrier, not the suitability of the habitat, has been the determining factor in the southern limit of the range of the species. The experiment was somewhat conservative, since the transplant was not the one that, in advance, would seem the most likely to succeed. That transplant would have been from the pop-

A predatory moth caterpillar eating an egg mass of Gillette's checkerspot near Rocky Mountain Biological Lab. The checkerspot was
not transplanted to an area free of its enemies.

ulation nearest to the barrier to the closest site on the unoccupied side
(where, for example, one might suppose the conditions at the two sites
would be most similar and the chance of establishment thus would be
at a maximum).

Nonetheless, as is characteristic of many, if not most, field experiments, the results do not constitute "proof." For example, it could be
that Gillette's checkerspot can never build *large* populations in Colorado
because conditions are so marginal there. And small populations, populations of, say, a few hundred individuals, are susceptible to accidental
extinction. Maybe individual female butterflies manage to cross the
barrier to Colorado every year and manage to establish small populations every hundred years or so. But perhaps those populations then go
extinct in a few dozen years. Such a pattern would be compatible with
the results of our experiments—since our successful transplanted population still is small enough that extinction is an ever-present possibility.
A more comprehensive test would require, among other things, establishing a series of transplant populations and monitoring them for a very
long time.

So some uncertainty about the answer to our original question inevitably remains. But the experiment is yielding other interesting results. The lack of a population explosion in the transplanted Gillette's checkerspots suggests that, unlike some invaders of new territory (such as *Opuntia* cactus in Australia), they have not been moved beyond the range of their natural enemies, or that they have been moved into a marginal rather than a super-suitable habitat, or both. The former hypothesis is supported by numerous observations of moth caterpillars, mites and bugs preying on egg masses and young larvae. The latter is suggested by the higher altitude of the Colorado population, which subjects the butterflies to more extreme climatic conditions. More important, in eight years the butterflies have not spread significantly out of the roughly 200-yard-square area into which the eggs and larvae were originally placed, even though apparently suitable habitat stretches in all directions. This dramatically demonstrates the sedentary habits of these butterflies and lends further support to the idea that failure to disperse and not lack of suitable habitat explained the absence of Gillette's checkerspot from Colorado.

Most biogeographic questions are not amenable to such experimental tests for logistic and ethical reasons. Often it is simply too complicated or expensive to make the transplant, or there are reasons to believe that the colonists might cause biological problems. The narrow range of host plants eaten by Gillette's and other checkerspot butterflies, the similarity and proximity of the habitats involved, and a variety of other factors made this transplant experiment safe. There seemed no possibility of the transplanted population becoming a pest; indeed if the experiment were successful, it would help secure the future of a species on the edge of endangerment. But we thought carefully and consulted widely before trying even such a restricted experiment. In relatively few cases are the requisite safeguards available to make such experiments both feasible and safe.

Biogeographers, as a result, are largely confined to analysis of the existing distributions of living and fossil organisms, combined with evidence of evolutionary relationships and the past physical configurations of habitats, in order to re-create the distributional history of Earth's biota. The seeming incongruity of some such lines of evidence is why the southern distribution of a variety of groups of plants and animals so puzzled biologists before the phenomenon of continental drift was accepted. Not only did biogeographers hypothesize that land bridges once connected the southern continents, but ornithologists were encouraged to assume (incorrectly) that the flightless birds were a polyphyletic group—that is, an assemblage of superficially similar species that independently lost the power to fly and were each more closely related to groups of flying birds than to each other. Such convergent evolution of flightlessness seemed a reasonable hypothesis when it was thought that the southern continents had always been where they are now and had at best been temporarily connected by land bridges. But now that continental drift is accepted, the descent of ostriches, rheas, and the like from a single flightless ancestor—as is indicated by their morphology—makes better sense.

Of course, there is some circularity in the reasoning back and forth between geology and biology—as is often the case for branches of science that deal with complex problems with a substantial historical component. The distributions of southern beeches, flightless birds, and *Lystrosaurus* fossils are explained by continental drift. But those distributions were also used as part of the *evidence* for continental drift. So was the existence of coal discovered in the Antarctic by Sir Ernest Shackleton's expedition in 1908. Its presence, along with fossils of the swamp-dwelling *Lystrosaurus* (its highly placed nostrils and other features have led to the surmise that it led a life similar to that of today's hippos), indicated that the climate of Antarctica must once have been warm enough to support a temperate biota—that it could not always have been located at the South Pole. This is evidence for drift, and it is part of the evidence that Pangaea moved northward before it broke up and the portion that is Antarctica returned south. But, in turn, drift helps "explain" the distribution of the coal and the *Lystrosaurus*.

The new geology, with its emphasis on plate tectonics and drift, has had a dramatic effect on biogeographic thought and has fueled the development of an extreme school of biogeography—vicariance biogeography. This school considers the development of barriers (by continental drift, changes in sea level, mountain building, desertification, and so on) within already-established broad ranges of organisms as virtually the sole explanation for present gaps in the distributions of related groups of organisms. It is the school's basic explanation for the existence of pairs of closely related species occurring in widely separated places (they call such pairs "vicariants," thus the name for the school). As indicated above, gaps in distributions caused by physical changes certainly do explain many biogeographic patterns—such as the widely separate locations of *Lystrosaurus* fossils. But many others, such as the establishment of a biota on the Galapagos clearly related to that of South America, are equally clearly the result of successful dispersal across barriers that have long been in place. The presence of the vicariant pair of the red admiral butterfly *(Vanessa atalanta)* in California and the Kamehameha butterfly *(Vanessa tameamea)* in Hawaii is also obviously due to dispersal. The geological evidence that both the Galapagos and Hawaiian Islands have never been anywhere near a mainland is overwhelming. The Kamehameha butterfly must be descended from a red-admiral-like ancestor that dispersed across the Pacific from North America, an explanation uncongenial to the vicariance biogeographers. In fact, many dispersal events have been observed—the repopulation of the island Krakatau by a wide variety of plants and animals after it blew up in 1883 being just one famous example. It is easy to see, then, why the vast majority of biogeographers assume that today's distributions are obviously the result both of the fragmentation of previously continuous ranges *and* of dispersal across established barriers.

Sorting out the processes of dispersal and range fragmentation and explaining phenomena such as "centers of origin" (geographic areas where there are many species of the same genus or many genera of the same family) are challenges for a revitalized field of biogeography. So is predicting the results of human interventions that could give organisms access to new habitats. There has already been a substantial debate on one possible such intervention—the creation of a sea-level canal across Central America. Such a canal would permit the distinct marine biotas on Atlantic and Pacific sides of the isthmus to invade each other, something that is not possible with the present raised, freshwater canal. (The issue is very complex, and while little in the way of consensus has been reached, almost everyone agrees that the results for one taxonomic group are likely to be quite different from those for another.)

But one of the most challenging areas of biogeography deals with far less grand events than the drift of continents or the artificial joining of two oceans. It deals instead with trying to figure out why, for example, one island has more species of birds on it than another, or why one species of plant is food for a greater variety of herbivores than another. And it is one of the areas of ecology where theory has greatly aided thinking about the problem.

The role of theory in science is poorly understood by the nonscientist—and small wonder. The word "theory" is used in several different ways. The scientific "theory" of evolution translates into the "fact" of evolution, in layperson's terms. That all living organisms are the modified descendants of other organisms fits all the known data, and is accepted by all knowledgeable scientists, even though, as you have seen, they often argue over details of the *mechanisms* of evolution. Other things that scientists also call theories are more tentative, however—such as the theory that the main role of plant secondary chemicals is defensive. That is a newer idea and has been less thoroughly tested; but it has been tested. The theory that one of a series of periodic comet showers exterminated the dinosaurs is a *hypothesis*. It is an explanation put forward to account for some observations, and clearly needs more testing (by gathering and analyzing astronomical and geological evidence) before it will be generally accepted or rejected.

When ecologists speak of "theory," as in the theory of island biogeography, they usually mean a relatively simple abstract framework that helps one to understand the complexity of nature. Very often the framework is constructed in the language of mathematics—as in predator-prey and competition theory—although in many cases it can be adequately described verbally as well. One such case is the theory of island biogeography. It was developed in the 1960s by two outstanding ecologists, Robert MacArthur of Princeton (who died tragically of cancer at the age of forty-two in 1972) and Ed Wilson of Harvard (the E. O. Wilson of *Sociobiology* fame).

Islands have long fascinated evolutionists in general and biogeographers in particular. One of Darwin's key observations dealt with islands:

> The most striking and important fact for us in regard to the inhabitants of islands, is their affinity to those of the nearest mainland, without being actually the same species. Numerous instances could be given of this fact. I will give only one, that of the Galapagos Archipelago, situated under the equator, between 500 and 600 miles from the shores of South America. Here almost every product of the land and water bears the unmistakeable stamp of the American continent . . . Why should this be so? Why should the species which are supposed to have been created in the Galapagos Archipelago, and nowhere else, bear so plain a stamp of affinity to those created in America?*

Why, Darwin wanted to know, if God created all beings at once, didn't island organisms everywhere more closely resemble one another than they resembled any mainland organisms? He pointed out that

* *Origin of Species*, First Edition, pp. 397–398.

. . . there is a considerable degree of resemblance in the volcanic nature of the soil, in climate, height, and size of the islands, between the Galapagos and the Cape de Verde Archipelagos: but what an entire and absolute difference in their inhabitants! The inhabitants of the Cape de Verde Islands are related to those of Africa, like those of the Galapagos to America. I believe this grand fact can receive no sort of explanation on the ordinary view of independent creation; whereas on the view here maintained, it is obvious that the Galapagos Islands would be likely to receive colonists, whether by occasional means of transport or by formerly continuous land, from America; and the Cape de Verde Islands from Africa; and that such colonists would be liable to modification;—the principle of inheritance still betraying their original birthplace.*

Indeed it *is* obvious that the organisms of the Galapagos are evolutionarily modified descendants of immigrants from South America, and the clear affinities of island biotas with those of adjacent mainlands is one of the many reasons that biologists accept the fact of evolution. Islands turn out to be wonderful places for the uninitiated to see evidence of evolution. For example, on the Galapagos, such animals as the land and marine iguanas show no fear of people—in striking contrast to the iguanas of the American mainland. The reason is clear; no terrestrial predators managed to disperse to the islands, and the predatory birds are too small to threaten these large lizards. Similarly, the native plants of the Hawaiian Islands are relatively defenseless. They generally are less poisonous to herbivores than their mainland relatives; Hawaiian mints, for example, lack the smelly leaves that indicate the presence of protective compounds. And Hawaiian raspberries are thornless. In both of these cases, large herbivores have been absent from the islands (until recently introduced by people), thus removing the selection pressure to maintain chemical or mechanical defenses against them.

Since populations on islands are always to one degree or another distinct from those on other islands or the mainland, they have, quite naturally, been "laboratories" for the study of speciation in isolation. But they also are where the obverse of speciation, extinction, has been most often observed. How could it be otherwise? Island populations tend to be small, and island organisms often have not had the opportu-

* *Origin of Species*, First Edition, pp. 398–399.

nity to coevolve with dangerous predators and close competitors. They are often easily eaten or outcompeted by tough new arrivals from the mainland.

So islands have long been of central interest to ecologists. But until the 1960s there was no coherent explanation of why some islands had many more species than others. Why, for example, does Cuba have many more species of reptiles and amphibians than the Caribbean island of Saba? Why are there many more species of birds on New Guinea than on Bali? One obvious "answer" is that Cuba and New Guinea are much larger than Saba or Bali. Cuba has almost 45,000 square miles; the little volcanic peak of Saba, less than 10. New Guinea has more than fifty times the area of Bali. But this "answer" is really just a version of a common observation—one that ecologists have formalized with mathematical treatment and that also seems intuitively obvious. This observation is that the number of species usually increases with the available space.

But this doesn't tell us *why* there are usually more species in a bigger area, nor does it explain the exceptions. For example, the Society Islands (Tahiti, Moorea, Bora Bora, etc.) have about the same area as the Louisiade Archipelago off the southeastern tip of New Guinea, yet the latter has several times the number of bird species. Indeed, the Hawaiian Islands are ten times as large as the Louisiades, and yet they have many fewer kinds of native birds. That complication is explained by the relative positions of the island groups. Both the Society and Hawaiian islands are much farther from "mainlands" than are the Louisiades. And it has long been noted that the number of species on islands generally declines with distance from the mainland. It was also long believed that, given enough time, this difference would disappear as more and more species found their way to the remote islands.

But MacArthur and Wilson took a somewhat different approach to explaining variations in species diversity on islands. In outline, their theory of island biogeography is simple. They proposed that the number of species found on any island would be determined by two opposing rates: that of immigration and that of extinction.

To help you see how this works, here is a simple example. Suppose a new volcanic island emerges from beneath the ocean off the coast of a continent inhabited by one hundred bird species. Mainland birds would soon begin to colonize the new island, but the *rate* at which new species would become established on the island would necessarily decline as the number of species there increases. The reason is straightforward; the number of new species that can invade from the mainland pool declines with each successful immigrant species. At the beginning, there are one hundred potential immigrants, so that a rate of one hundred successful immigrations per year is theoretically possible. As soon as one species becomes established, the highest possible immigration rate for the succeeding year drops to ninety-nine. At the end the highest possible annual rate is zero, because all of the species will have made it over.

On the other side of the process, the rate at which species go extinct on the island will also be a function of the number of species already there. When the island is almost empty, that rate will be very low, because there are few species present that can become extinct. When the island is inhabited by the highest possible number of species, the extinction rate will be at its highest, because there are then more candidates for extinction. Also since the island has finite resources, the more species are present, the smaller their individual populations are likely to be, and therefore the more likely they are to die out.

In short, when the island is relatively empty, the immigration rate would be high and the extinction rate low. The reverse would be true when the island was relatively full. Accordingly, at some point between an island population of zero species and one of one hundred species, the two rates must be equal. Input (immigration) would balance output (extinction). At that point, the number of species on the island would remain constant as long as the factors that determined the rates of immigration and extinction did not change. That number is the equilibrium number of species on the island. To put it another way, the number in the cast would remain the same, but because there would be a steady turnover of species as some went extinct and others replaced them, the actors themselves would gradually change. The idea of a balance of diversity created by opposing rates of immigration and extinction is the kernel of MacArthur and Wilson's equilibrium theory of island biogeography.

The theory has a number of interesting corollaries. Everything else being equal, islands more distant from the mainland will have lower immigration rates. Fewer species will arrive per unit time even when the island is almost empty. The lower immigration rate will be counterbalanced at a lower extinction rate, such as is found when there are relatively few species on the island. Consequently, the input and output rates will balance at a lower level of species diversity on the island. Similar reasoning leads to the prediction that islands near the mainland will have a greater number of species because they will have higher immigration rates, and smaller islands (with higher extinction rates) will have fewer species than larger islands—always with everything else being equal.

You should note that the equilibrium theory does not take into account interactions among the species. That is, one of its simplifying assumptions is that competitive, predator-prey, and mutualistic relations do not significantly influence rates of either immigration or extinction. Obviously, this assumption is sometimes, perhaps often, violated —but it is precisely the kind of simplification that makes mathematical theories useful tools for thinking about (rather than completely describing) the natural world.

We can look to the volcanic explosion of the island Krakatau, between Java and Sumatra, on August 27, 1883, to find one famous example that supports the equilibrium theory. As a result of the violent eruption, the flora and fauna of the small Krakatau island group was totally destroyed. Indeed, windows one hundred miles away were shattered, and the resultant tsunamis (tidal waves) killed some 36,000 people on nearby islands. Subsequent biological surveys showed that fourteen species of birds had repopulated the Krakatau group by 1908, sixty were present by 1919–1921, and sixty-four by 1932–1934. Not only did the *rate* of increase in species diversity decline as predicted by the theory, but as the islands approached equilibrium, some turnover occurred as well. During the fifty years of these censuses, seventy-one bird species actually became established, but seven of those again went extinct.

E. O. Wilson and his student Dan Simberloff put the theory of island biogeography to another test in one of the most famous series of field experiments ever done in ecology. They hired an exterminator to put tents over entire (though quite small) mangrove islets in the Florida Keys and, from the point of view of the insect inhabitants, created mini-Krakataus with methyl bromide gas. Then they carefully monitored the defaunated islands (the mangroves and other plants were unharmed) and recorded the return of insects, spiders, mites, and other arthropods. It took less than a year for most of the islands to return to their original species diversity—and they did so in an interesting way. They tended to overshoot the original number of species and then decline to an equilibrium value. The results of these island experiments conformed to some of the most important predictions from the theory.

The theory of island biogeography has been considerably elaborated beyond the simple formulation presented here. There has been research on the shape of the curves that plot the two rates against the species diversity of the island and what happens when islands are clustered closely together. Island biogeographic theory has also been applied widely to problems ranging from attempts to understand the diversity of herbivorous insect species eating a given plant (considering the plant as an "island") to consideration of the faunas of entire continents, in which speciation rates substitute for immigration rates and their current diversity is viewed as representing an equilibrium between speciation and extinction.

Some of the most interesting applications have been in the burgeoning new subdiscipline of ecology known as conservation biology, especially in connection with a problem now known as "insularization." As growing human populations and activities destroy habitats, the remaining areas occupied by more or less unperturbed communities become increasingly like islands in a sea of disturbance. For several reasons, this insularization leads inevitably to a loss of diversity. Suppose that several small reserves were set aside in a large area of forest, and that later the parts of the forest outside of the reserves were felled. Some species would be lost simply by accident because at the time of the clearing no individuals happened to be in the reserves. Others would be lost because a vital prey species or a keystone mutualist was accidentally excluded, or because the reserve was simply not large enough to accommodate their territorial or other resource requirements.

But, over time, the reserve would probably also lose some of the species that were originally present as viable populations. The reason is explained by island biogeographic theory. When the forest was intact, migration of individuals from place to place within it was relatively unconstrained. But when the forest was fragmented, migration between the resulting isolated reserves would be much less frequent. Some species would be physically unable to cross the barriers of nonwooded land; others, including many birds and butterflies that are physically capable of crossing, simply would not do so. So migration into reserves would be much reduced once they were isolated.

In contrast, extinction rates would rise. Population sizes of many species would be smaller after isolation because the amount of suitable habitat for them would be reduced. That is because beforehand the populations of those species had occupied habitat both inside and outside of the reserves, or because a portion of a reserve, once entirely embedded in a larger forest, now would be forest edge—an unsuitable habitat for many woodland species. And, as you recall, smaller populations are more vulnerable to extinction. According to the equilibrium theory, if either immigration rates are reduced or extinction rates are raised, the equilibrium number of species will fall, or "relax," to a new, lower equilibrium. When both occur simultaneously, the number of species present will fall even further—for animals, there may be what is known as a faunal collapse.

Unfortunately, this is not all just theory. One of my first graduate students, Michael Soulé, and one of his students, Bruce Wilcox, have been prime movers in the field of conservation biology (Bruce is now executive director of Stanford's Center for Conservation Biology) and have been deeply involved in relating faunal collapse to the problems of designing nature reserves. In order to understand the basic process, they analyzed the relaxation of the land mammal fauna of the Sunda shelf of Southeast Asia. As the Pleistocene glaciers melted about 10,000 years ago, rising sea levels fragmented the mainland on the shelf into Borneo, Sumatra, Java, and many smaller islands, all isolated from the Malaysian mainland to which they had previously been connected.

Fossil evidence indicates that, at one time, each of the larger islands had essentially the same fauna as the mainland, and today Sumatra, Java, and Borneo *combined* still do. But each island is now missing some species. Sumatra lacks the banteng (a wild ox) and the leopard; while six species, the Sumatran rhino, Indian elephant, Malay tapir, Malay bear, Simiang gibbon, and orangutan, all found on Sumatra, are absent from Java. The Javan rhino, Malay tapir, tiger, leopard, and gibbon are all missing on Borneo. These absences indicate that the fauna of each island has significantly "relaxed" since it became isolated; lower immigration rates and higher extinction rates have lowered the equilibrium number of species.

Mike, Bruce, and a colleague, Claire Holtby, have used this information about the fate of the mammal fauna in the naturally formed island "reserves" of the Sunda shelf to predict the probable fate of large mammals now being "conserved" in Africa. They applied the relationships between decline in species diversity and island size on the Sunda shelf to nineteen East African national parks and game reserves and predicted the course of faunal relaxation there. Their conclusions were not encouraging. After it becomes completely isolated from other natural habitat, a 2,500-square-mile reserve is predicted to lose 11 percent of its fauna in 50 years, 44 percent in 500 years, and 77 percent in 5,000 years. Smaller parks suffer even more dramatic losses.

There are, needless to say, many uncertainties in these estimates, but they are the best produced so far. Indeed, other research indicates that many populations of large animals (megafauna) in African (and American) national parks are too small to have much chance of long-term survival. This means that even if preserves are flawlessly guarded, much of the spectacular African fauna will disappear in the not too distant future. On the positive side, people could move animals from park to park, re-establishing some that had gone extinct in one with immigrants from another. This would postpone the extinction process —although it would probably further limit the possibilities for future evolution of the fauna by slowing the rate of speciation. On the negative side, of course, most of the African parks are already threatened with poaching, intrusion, and utter destruction as surrounding human populations grow at rates of 3 to 4 percent per year—so the extinction of the African megafauna may occur much more rapidly than the theory predicts.

But more is at stake than that spectacular African fauna. A very important question for ecologists in particular and humanity in general is how rapidly the enormous biological riches of tropical rain forests will disappear as those forests are increasingly fragmented. The answer to that question is being sought directly by Thomas Lovejoy, scientific director of the World Wildlife Fund (WWF), who has organized an impressive island biogeographic experiment near Manaus, Brazil. The experiment is a joint project of WWF and Brazil's National Institute for Amazonian Research (INPA). The "Minimum Critical Size of Ecosystems" project is one of the most ambitious ecological investigations ever launched, and also one of the most important. Among other things, it is censusing the various communities in rain-forest patches before and after they are isolated by deforestation, gathering a set of data that not only will test various aspects of the equilibrium theory but also be invaluable in providing guidelines for the establishment of forest reserves. Special attention is being paid to birds, mammals, butterflies, ants, and plants.

The Brazilian-American team, headed by Tom and Herbert Schubart of INPA, has gained the cooperation of some large Brazilian cattle-raising corporations, which by law are required to leave a portion of their land in forest as they clear range. Tom has persuaded them to adjust the areas left undisturbed to form a series of square patches ranging in size from 2.5 acres to about 20,000 acres (about 30 square miles). In most size classes, more than one "reserve" will be left, so information can be gathered on the consistency of the fate of faunas in the same-sized forest remnants.

The logistics of the project have been daunting. It has required coordinating with the cattle companies, and feeding, housing, and transporting dozens of workers. People doing research on the project have had to contend with a nasty protozoan disease (American cutaneous leishmaniasis), accidents, and at least one scary but ultimately harmless encounter with a pair of jaguars with a cub.

In spite of the problems, progress has been substantial—and results are already being obtained after only a few years, although answers to questions relating to the equilibrium theory are still far in the future. Preliminary studies indicate that mixed-species flocks of insectivorous bird species, an important part of Amazonian bird communities, are especially sensitive to forest fragmentation. Reserves will presumably have to be quite large to protect them. The problems of preserving the Amazon forest may seem far away to most Americans. But concern has been growing among biologists that declines observed recently in populations of certain migratory North American songbirds are due in part to destruction of habitat in the Latin-American tropics. The birds spend the winter there as members of just the kinds of communities the INPA-WWF team is studying.

Our Center for Conservation Biology at Stanford has been conducting a study, under the leadership of Bruce Wilcox, with goals similar to the INPA-WWF Amazon project, but with a different design and in a totally different environment. Instead of the biogeography of tropical-forest fragments, we have been investigating the island biogeography of butterflies in the mountains in the Great Basin—an area, mostly in Nevada and Utah, that, unlike most of the rest of the United States, has no drainage to the sea. The project was suggested to us by declines in the size of the Jasper Ridge demographic units of the Bay checkerspot during the California drought. Those declines made it seem likely that the Jasper Ridge Preserve was too small to support that species over the long term. The apparent peril of the Bay checkerspot impressed upon us the need both for more information on insect biogeography and for more information pertinent to the design of reserves. How else could progress be made toward the preservation of elements of Earth's biota that are less spectacular than elephants and antelopes—elements including insects and many other invertebrates, herbaceous plants, and so on?

The Great Basin is an immense plateau from which more than two hundred isolated mountain ranges rise. Some thirty of these have peaks reaching above 9,000 feet. The floor of the basin at 5,000–6,000 feet is desert, but above 7,500 feet or so, the ranges support areas of meadow and woodland—essentially "islands" of relatively moist habitat in a "sea" of desert. There is abundant evidence, however, that the situation was not always thus. Roughly 10,000 years ago, when much of Earth's water was still tied up in glaciers (and sea level was sufficiently low that today's islands of Southeast Asia's Sunda shelf were connected to the mainland), the climate of the entire Great Basin was cooler and wetter, and moist habitat was ubiquitous. Then, as the water melting off the glaciers caused the sea level to rise, the Great Basin climate became drier, producing the habitat "islands" on the mountain ranges—a process analogous to the fragmentation of previously continuous habitat by human activities.

The mountain-range islands are of many different sizes and many different distances from the two "mainlands," the Rocky Mountains and the Sierra Nevada. Fortunately for the project, the biogeography of mammals and birds of some of the ranges had been studied earlier, providing an instant comparison with our studies of butterflies and plants. We gathered our data in four summers by keeping an average of three research teams (three to five people each) in the basin each season. They had no problems with leishmaniasis or jaguars but did suffer infrequent gut upsets (some strange parasites occur in the basin waters) and occasional rattlesnake sightings.

Previous work by James Brown of the University of Arizona at Tucson had indicated that the small mammal faunas of the ranges were not in equilibrium, but rather, even after thousands of years of isolation, were still relaxing. The "islands" are so isolated that their immigration rates are almost nil, and as a result their small mammal communities are decaying away toward a species diversity of zero. As one would expect (because extinction occurs much more frequently in smaller islands), the smaller mountain ranges have very few mammal species left, while the large ones have considerably more.

Birds, in contrast, immigrate to the islands at a very high rate, and it is not surprising that very small mountain ranges have almost as many species as very large ones since species that go extinct are quickly replaced. The bird fauna has thus not suffered a decline as severe as that in the small mammal fauna. But as Brown has pointed out, the equilibria of the bird populations appear somewhat different from that envisioned in the theory. Rather than the successful immigrants being a random selection from the available pool, in most cases they are probably members of the same species that just went extinct, arriving and fitting into the habitat and role of their predecessors. Each range is almost continually bombarded with avian immigrants representing the entire mainland pool, keeping all the habitats of each range more or less saturated with the bird species that fit into them. In short, the fragmentation of the once continuous moist habitat of the Great Basin has had less impact on the species diversity of birds than small mammals, because the fragments are (for birds) not so isolated.

Our preliminary results show the pattern of species diversity in Great Basin butterflies to be similar to that already found in birds, although it seems that the reasons are somewhat different. We think that immigration rates are lower for butterflies than for birds, but that extinction rates are also lower. Butterfly equilibria in the Basin appear to be more like those described by island biogeographic theory than bird equilibria there—with more isolated islands having a lower immigration rate than less isolated ones. The results suggest that isolated reserves will lose their butterfly diversity more slowly than they will lose that of mammals or birds (butterfly immigration rates are much higher than the mammals', and their extinction rates are much lower than the birds'), but they will inevitably lose it.

In order to gain more insight into what controls the numbers of butterfly species in a given mountain range, we are now studying the distributions of different kinds of habitats and larval food plants *within* ranges, especially between the valleys of different streams in the same range, valleys that appear to be analogous to small islands in archipelagoes and thus amenable to analysis against a background of the equilibrium theory.

Island biogeographic theory has provided the basic stimulus for these and many other studies. But the theory, as one might expect, has also been controversial. In fact, one of its critics has been Dan Simberloff himself, who with Ed Wilson did the pioneering work on the biogeography of mangrove islets. Among other things, researchers have questioned whether there is really time for species numbers to tend toward an equilibrium. In many situations, the rate of turnover may be so low that there is simply a general trend of increase or decrease in species diversity over geological time. In many cases, the theory is difficult to apply because the mainland source of the biota is hard to specify (this is a greater problem in archipelagoes than in single islands off continental coasts). And the area of an island, which under the theory is the measure of its carrying capacity, is at best an indirect measure.

Largely because of this last point, Simberloff has questioned the application of the equilibrium theory to problems in conservation biology. He suggests that the details of habitats and biotas usually override simple area effects, and are the critical factors in determining what species can be preserved in which reserve. For instance, the theory generally supports the view that reserves should be as large as possible, everything else being equal. But no matter how large a reserve is, if it doesn't include populations of the proper food plants, it will not support a population of the Bay checkerspot, whereas a relatively small reserve with those plants might do so. But, of course, the key is the "everything else being equal"—and I think with that constraint in mind, the theoretical conclusion that reserves should be as large as possible is valid.

An early critic of the equilibrium theory was British evolutionist David Lack, who became famous for his pioneering studies of Darwin's finches on the Galapagos. Lack disagreed with MacArthur on why distant islands had impoverished plant and animal communities. Lack thought that the paucity of competitors on islands favored the evolution of communities of relatively unspecialized species (those that could persist in a wide range of physical conditions and use a variety of resources). He thought that such species were able to resist invasion by more specialized species. He felt that birds, at least, can reach islands about as easily as they reach other habitat patches, that distance from the mainland only affected the fauna in a minor way, and that extinctions were quite rare.

MacArthur agreed with Lack that the ecologically impoverished communities of distant islands resisted invasion, but he had less faith in the ease of dispersal. He held that the low number of species on distant islands was explained principally by the equilibrium theory, and that distance from the mainland was a very important element of that.

Regardless of the eventual resolution of these controversies,

MacArthur and Wilson's theory can serve as an example of the utility of good mathematical theory in ecology. Its value was expressed very well by Lack himself, who wrote in the first chapter of *Island Biology* (University of California Press, 1976): ". . . while I here differ fundamentally from MacArthur and Wilson, their highly original re-appraisal of 'island biogeography' revitalized a supposedly dead subject, and I have reread their short book of 1967 more times than any other on biology, because I keep finding further new and revealing biological insights in it."

Biogeography deals mostly with the past and present distribution of life-forms; but, as we have seen, it can also speak to future distributions (for example, in the form of island biogeographic predictions about relaxation). Indeed, from what is known about ecology in general and biogeography in particular, one can make some rather firm predictions about the nature of the biota in the future. For instance, unless current trends are altered, Earth's biota will soon be much less diverse than it is now; and by 2100 more than half of the species of plants and animals that exist today will be extinct. The consequences for humanity of such an epidemic of extinctions may be catastrophic. This loss of diversity will not affect all communities equally; the tropics, with their disproportionate richness, will suffer disproportionate losses.

A second prediction is that if civilization is able to survive in the face of those losses and the difficulties associated with them, it seems certain that by the end of the next century very few communities will be "natural" in any current sense of the word. Much more of the planet will inevitably have been converted to simple agricultural ecosystems in the struggle to feed humanity's rapidly growing populations. One result of this is that, increasingly, *Homo sapiens* will actively have to manage the biosphere to preserve what diversity is left; wilderness will be "replaced" by reserves, parks, zoos, and botanical gardens. Whether ecologists will have learned enough by then about the properties of natural communities to allow them to manage unnatural communities successfully remains to be seen.

WHO LIVES TOGETHER,
AND HOW

Community Ecology

IN THE northern part of the Australian Great Barrier reefs, as many as 2,000 species of fishes live together. In a few square miles of Brazilian rain forests, 700 kinds of butterflies (about the number that occur in all of North America north of Mexico) may fly among hundreds of species of trees. Some 1,400 species of birds breed in Colombia, which has an area less than one-sixteenth that of North America (which has only about 650 species of breeding birds). The Serengeti National Park of Tanzania supports 24 different species of antelopes, gazelles, and buffalo (all members of the same taxonomic family as domestic cattle), and almost 400 distinct kinds of birds brighten the landscape. As many as 30 million species of arthropods may creep, hop, and fly through the tropics—perhaps a quarter of them beetles! (When a theologian asked the great English biologist J. B. S. Haldane what could be concluded about the Creator from studies of creation, Haldane is reputed to have answered that "He must have had an inordinate fondness for beetles.") And though not as diverse as animals, or even beetles, flowering plants have more than a quarter million named species. In short, a multitude of other kinds of organisms shares Earth with us, and the diversity of species living together varies greatly from area to area.

Finding explanations for patterns of diversity (that is, varying numbers of different species) in natural communities has always been a daunting challenge for ecologists. Island biogeographic theory just scratches the surface of the problem. It cannot deal with many of the questions that diversity poses: Why should there be a different number of warbler species in two one-acre patches of woods that are the same distance from the forest? Why are there many more kinds of butterflies and trees (indeed, of most kinds of organisms) in the tropics than in temperate zones? Why are there so many species of beetles? And can species that play exactly the same role coexist in a single community more or less permanently?

We'll see that partial answers have been found to some of these questions. In the untangling of these complexities of community ecology, the work of the late Robert MacArthur stands out. He left an indelible mark on this subdivision of ecology, the one that tries to account for the particular combinations of species found living together in various habitats.

In the 1950s, the field of community ecology was in the doldrums. Ecologists knew that the complexity of communities varied from one place to another. Some communities contained great numbers of species; others, relatively few. They also knew that after disturbance, the original community often tended to restore itself—if a forest were chopped down and converted to a farm field, and then if the field were neglected, usually a forest very similar to the original would eventually grow back. In addition, work in the first half of this century by pioneer ecologists such as England's Charles Elton had revealed some general community patterns—for example, that particular sorts of organisms were found in the biological communities of particular climatic zones. An enormous amount of literature *describing* communities had accumulated, competition between plants had been documented, and naturalists often had an excellent "feel" for how organisms were associated with one another. But quantitative and repeatable methods did not exist for predicting how many and which kinds of organisms would be found in association. At midcentury, there was a lack of new hypotheses and tests of hypotheses, and, as a result, little sense of excitement in community ecology.

Then along came MacArthur with a series of stimulating and controversial ideas. Among them, he proposed a simple model to describe the patterns of relative abundance of bird species; he developed a theory to explain how many species could subsist on a given set of resources; and he examined the interconnections of food chains to see if they could provide a basis for predicting the stability of communities. In all cases, as in the theory of island biogeography (which actually came rather late in his career), MacArthur tried to cut through the complexity of biological communities to find underlying rules that explained their properties.

Community ecology has not been the same since. MacArthur quickly acquired adherents, who continued investigations in his style, and critics, who claimed (among other things) that he was oversimplifying. The argument continues to this day, as already indicated by Simberloff's disenchantment with island biogeographic theory. Students who only know of Robert MacArthur from textbooks of ecology tend to think of him as a mathematician. Those of us fortunate enough to have known him personally are more inclined to think of him as a naturalist and bird watcher with an extraordinary talent for gaining insights into complex systems. Sometimes his insights were (even in his own later view) dead wrong. But the very way he did ecology, even more than his elegant and often robust results, was his central contribution.

Although community ecology was in the doldrums in the 1950s, population ecology (which deals with populations of single species) was not—and a foundation at the population level is a prerequisite for erecting an edifice of community ecology. Then (as now), explaining why each species of plant and animal lives where it does, and why individuals of some species are more numerous than others, were seen as fundamental tasks of ecology. Indeed, the classic 1954 work by Australians H. G. Andrewartha and Charles Birch on population ecology was entitled *The Distribution and Abundance of Animals* (Chicago University Press).

You now know that some of the explanations for the distribution and abundance of organisms lie in the physical heterogeneity of Earth's environments and in their history. Fishes are not found far from water, mammals with heavy coats of white fur and insulating layers of fat under their skins are not denizens of tropical rain forests, and penguins are still confined to the cold-water parts of the Southern Hemisphere. If that were all there was to it, the task of community ecologists would be simple indeed—just an amalgam of the work of physiological ecologists (to explain which organisms can live in a given set of physical conditions) and historical biogeographers (to determine which organisms had access to places having those conditions).

But things are not that simple. In their book, Andrewartha and Birch identified four components of an animal's environment—weather, available nutrition, a place to live, and other organisms of different kinds. The first three have already been discussed, and in any case their importance is intuitively obvious. It is the last that causes complications for community ecologists. The influence of organisms on the distribution and abundance of other organisms was and still is most controversial. Can a predator influence the number of species of available prey? Can one species of organism exclude another from a geographic area by outcompeting it? What does determine who lives where?

A version of that last question was one of the first to interest MacArthur: what determines the diversity of bird species in a given habitat? MacArthur knew that bird watchers acquire a "sixth sense" about what kinds of birds will be found in a given chunk of habitat, but they can't explain how they know. He wanted to discover some measurement that could be made in a meadow or patch of woods that would in a limited way duplicate what most bird watchers could do without thinking: predict the number of bird species living there. The way MacArthur tackled the problem exemplifies his approach to nature. He reasoned that if a bird watcher could tell with a glance that ovenbirds or northern parulas would be found in a certain patch of woods, obviously they were using some gross characteristic of the woods for prediction. He set out to discover what that characteristic was and to measure it.

MacArthur, using his own bird watcher's intuition, made an initial guess that the vertical pattern of the vegetation was important. After all, birds are often thought of as either occurring typically in open fields or in woodland, and the most obvious difference between the two is that the vegetation is a lot taller in one than the other. Could finer differences in the vertical pattern also be important? In order to find out, MacArthur developed a quantitative index of the vertical structure of the vegetation in a habitat, which he called foliage height diversity. That index was a simple measure of the variation in the proportion of foliage occurring at different heights above the ground.

To calculate it, all one needs is a measuring tape and a white board that can be mounted at different heights on a pole so that the proportion of it obscured by vegetation can be seen. The details of the procedure need not concern us, but it yields a single number that is low if all of the foliage is found at a single level regardless of height (as in a treeless field of grass or a woodland with no undergrowth) and high if about the same amount of vegetation is found at several different heights above the ground (herbs, bushes, and trees growing intermixed).

MacArthur calculated this index for a series of habitats. He also did bird censuses in the same habitats, and compared a statistic that measures diversity of bird species with the one measuring foliage height diversity. He found a general principle: bird species diversity in decid-uous forests could be accurately predicted from a knowledge of foliage height diversity. Curiously, the diversity of plant species did not itself influence how many bird species were present—except to the extent that it influenced the foliage height diversity. It did not matter, for example, whether a dense canopy above twenty-five feet was formed by one or five species of trees. But it did matter if those trees had so blocked out the light that no bushes could grow in the two-to-twenty-five-foot range. By this study, MacArthur demonstrated that Andrewartha and Birch's "other organisms of different kinds" (in this case, plants) could have a dramatic influence on the distribution and abundance of birds —and that the plants do it primarily through their structural arrange-ment, rather than through their species diversity. Thus part of the an-swer to MacArthur's original question of what determines bird species diversity is "foliage height diversity."

MacArthur's study illuminated a general principle underlying the apparent enormous complexity seen in most communities. One could easily imagine that many complex interacting factors would all be im-portant in determining bird diversity; MacArthur found that a single, easily quantifiable one dominated. A cynic might argue, however, that the careful MacArthurian measurer could only do laboriously what a first-rate birder could do at a glance. But he would be missing the point: the birders never discovered the general principle. Furthermore, the birder couldn't tell an inexperienced person how he or she predicted how many species there would be in a given area; MacArthur could. Results must be communicable and communicated; otherwise they aren't science.

Besides being theoretically important, MacArthur's principle has practical significance. If a timber company familiar with the principle wished to replant a logged area in a way that would enhance bird diversity, it would know that planting to provide understory vegetation (layers of plants beneath the forest canopy and shaded by it) as well as large trees would likely be more important than how many tree species were planted. Or if the Nature Conservancy were faced with a quick decision about which of two pieces of wooded land to purchase as a bird sanctuary, it now has available to it a quick way of ascertaining which will support the greater diversity of birds.

Understanding species diversity has grown increasingly important since MacArthur's work on foliage height and birds. Many more people are now aware that diversity is among the most precious nonrenewable resources of our planet. That awareness has grown in part because organic diversity is now gravely threatened around the world by human modification of environments with tools as varied as plows, bulldozers, acid rain, and possibly thermonuclear weapons.

So we come to a fundamental question about this great but threatened resource of diversity: why does it exist? Why are there so many different species of organisms? That question, as a focus for ecologists, traces to MacArthur's renowned teacher, G. Evelyn Hutchinson of Yale University. More than a quarter of a century ago, Hutchinson was in Sicily looking for some long-lost species of water bugs. Near the cave where the bones of Santa Rosalia, the Patron Saint of Palermo, had been discovered, Hutchinson found a pool rich in bugs of two species of the genus *Corixa*—one large, and one small. It started him wondering why there were "two and not 20 or 200 species of the genus in the pond," a question that led him to ponder the central mystery of ecological diversity. He wrote about possible solutions to that mystery in a classic article entitled, "Homage to Santa Rosalia, or why are there so many kinds of animals?"

Hutchinson considered several possible reasons for the great variety of animal species. He noted that the complex three-dimensional structures created by plants, as they invaded the land and competed for light, provided a variety of *niches* for different kinds of animals, such as insects, to occupy. The niche of an organism can be defined loosely as its "way of life," its "occupation," or its method of acquiring resources. An insect species can find a niche in a tree by boring into its trunk, eating the wood as it goes, by consuming its leaves, by sucking its sap, by drinking nectar from its flowers, or by eating the tree's borers, leaf-eaters, sap-suckers, or nectivores. The more such niches there are, the more insect species there can be.

But how many species of leaf-eaters or sap-suckers can live together on the same species of tree in the same area? That is, how many functionally similar species (members of the same guild) can there be within a single community?* Is there something that keeps the number of different species in a guild from increasing without limit? Hutchinson thought that there were definite limits to how similar two species could be and still co-occur. In fact, he suggested that in many cases a series of coexisting guild members would differ in size by a predetermined amount. Such size differences might be an indication, for example, of the size of prey that each animal could use. If the prey taken by the predators were the same size, that would mean the two predator species occupied the identical niche. Therefore there should be a "limiting similarity" for coexisting species. Species more similar than that could not live together—one would outcompete the other and exclude it from the area.

A basic tenet of ecology has long been that two species could not coexist if their requirements were too similar—more explicitly, if both depended on one resource that was in short supply. This idea of competitive exclusion has been called Gause's principle, in honor of G. F. Gause, who, in addition to his work on predator-prey systems, did early laboratory experiments on interspecific competition (that occurring between species), using microorganisms. Gause's experiments were precursors of the sort of work that Park did with flour beetles. Thanks in large part to Robert MacArthur, Gause's principle has been expanded into a body of mathematical theory that predicts the precise conditions under which species can or cannot coexist in nature, what the limiting similarities are (how much alike they can be), and thus under what circumstances the presence of one or more species can prevent other species from invading an area.

In a famous paper (in fact, his doctoral dissertation), MacArthur showed that five closely similar insect-eating warblers that lived in the same area behaved "in such a way as to be exposed to different kinds of food." For instance, the blackburnian warbler fed mostly at the ends and middles of upper branches, while the bay-breasted fed, on the average, lower in the tree and more toward the base of the branches. These observations conformed to ideas of competitive exclusion and

* "Community" is often used to mean all the organisms in an area, but it also can be used in more restricted senses as in "bird community" or "plant community" or "ambush predator community." In the latter usage, it would be synonymous with "ambush predator guild."

limiting similarity—that closely related species either did not occur together or, when they did, showed distinct differences in sizes or their
ways of life. Often, as in the warblers, they used different resources.
Such "resource partitioning" abounds in other groups—Edith's and
chalcedon checkerspots at Jasper Ridge feed as caterpillars on entirely
different food plants, for example.

The related ideas of niche, competitive exclusion, and limiting similarity are critical to one modern view of what controls diversity in
biological communities (we'll deal with another later). In this view,
biological interactions are considered to dominate the evolution of communities and to limit the number of different kinds of similar organisms
that can be found in them. Plants, animals, and microorganisms living
together have influenced the sizes of one another's populations and
coevolved to produce the composition of modern communities. And, of
these dynamic and coevolutionary interactions, competition is one of
the most important.

The centerpiece of this view is a body of mathematical ideas known
as "niche theory." It, too, traces to Evelyn Hutchinson, particularly to
an influential piece that he published in 1957. In that paper, Hutchinson developed a multidimensional geometric definition of the niche of a
species. One dimension of a wren's niche might be temperature, with
the bird being able to survive within a certain range of temperatures.
The wren would die of overheating above that temperature range or of
cold below it. Another dimension might be size of nest cavity, again
with a range restricting that dimension of its niche—with too small a
hole, the wren can't get in; with too large a hole, egg predators can
enter. A third dimension might be the size of insect prey: suitable prey
for this wren would lie between those too small and those too large to
be efficiently eaten.

Hutchinsonian niches can be a little complicated, since they may
be defined with more than three dimensions and thus can involve abstract volumes that mathematicians call "hypervolumes." But in practice, niche theory most often deals with only one dimension, usually
that defining a key limiting resource. Thus, for example, the possible
competitive relationships of populations of two species of wrens might
be studied by comparing their use of insects. The result would be curves
plotted over the insect-size axis (a line which represents the insect
resource dimension, with the size of the insects increasing from left to
right) that show what proportion of each wren population feeds on

different-sized insects. These curves might be superimposed on one another (resource totally shared), might partially overlap (some resource partitioning), or might show no overlap (complete resource partitioning).

The view that interactions, especially competitive interactions, were central to understanding community structure and the limits to diversity has naturally led to studies of sets of species that seem to have closely similar niches. During the middle years of the twentieth century, numerous studies revealed a common pattern in closely related organisms —occupancy of adjacent but more or less exclusive niches (ones with little or no overlap on one—or more—axes). For instance, in open marshes of the U.S. Pacific Northwest, yellow-headed blackbirds establish territories and nest in cattails and other vegetation emerging from relatively deep water. They hold the most productive territories (those that produce the most insect food) on the most productive lakes. Male yellow-headed blackbirds have little luck in attracting females to the poorer territories near the shore. Red-winged blackbirds, in contrast, are confined to shallower areas of the marshes or to drier locations. The redwings normally arrive at the breeding grounds before the yellow-headed blackbirds and occupy rich territories in the vegetation emergent from the deep water. But when the yellow-headed blackbirds arrive, they displace the redwings.

Yellowhead pairs are less successful at exploiting resources in the less productive sites, but in these places the redwings can still thrive. That the former have a larger body size is possibly a result of selection to permit them to oust the redwings from the richer territories, or possibly because it permits them to eat larger seeds where they overwinter (or perhaps it is a combination of the two). At any rate, the utilization curve of the redwings clearly covers more of the "habitat richness" axis than that of the yellowheads, since they can survive under a wider range of conditions. But the yellowheads, with a narrower niche, normally exclude the redwings from the yellowhead niche. As a result, the redwings are said to have a narrower "realized niche" (what they actually occupy) than fundamental niche (what they could use in the absence of their competitor).

My Stanford colleague Craig Heller studied a similar situation in chipmunks of the California Sierra Nevada. Four chipmunk species, the least *(Eutamias minimus),* yellow-pine *(E. amoenus),* San Bernardino *(E. speciosus),* and alpine *(E. alpinus),* occur in four contiguous altitu-

CRAIG HELLER, STANFORD UNIVERSITY

The alpine chipmunk seems to be limited to the alpine zone of the Sierra Nevada by the aggressive behavior of the San Bernardino chipmunk that occupies the lodgepole-pine zone below it.

dinal zones—sagebrush, piñon pine–sagebrush, lodgepole pine, and alpine respectively. Heller showed in physiological experiments that the fundamental niches of the four species overlapped considerably, even though their realized niches were decidedly different.

One factor in the zonation was straightforward. The desert habitat of the least chipmunk was apparently too hot for the other three to invade—it was outside their fundamental niche. In laboratory experiments, Heller found patterns of interspecific aggression that further explained the mechanisms underlying the altitudinal habitat partitioning. Two chipmunks seem to hold aggressive dominance over the alpine chipmunk: the San Bernardino in the habitat above it in altitude, and the yellow-pine chipmunk below it. Similarly, aggression by the yellow-pine apparently keeps the least chipmunk from spreading to higher altitudes; in areas lacking the yellow-pine, the ecological range of the least chipmunk extends upward. Why the San Bernardino chipmunk has not extended its range upward and downward is not clear, but Heller suspects that it is due to subtle physiological stresses on this large-bodied species in the more arid piñon pine–sagebrush habitat below and the alpine habitat above—stresses not revealed by short-term laboratory experiments.

In an investigation of another set of closely related species, Earl Werner of Michigan State University compared the sizes and diets of two freshwater fishes, the small bluegill and the bigger largemouth bass. He found that the prey taken were of quite similar sizes—the positions of the niches of the two fishes were quite close to each other on the resource axis. Werner thus used the theory of limiting similarity to predict that an intermediate-sized species, the green sunfish, could not coexist with the bluegill and largemouth bass. This species would take prey on the average too close in size to those taken by either of the other two—it would compete too much. His prediction was tested with the use of scuba gear. Although all three species live in the same lakes, underwater observations confirmed the prediction from theory that they do not live together. The bluegills and bass dwell in the open water; the medium-sized green sunfishes live in the weeds.

Many other pairs of species with closely similar niches have exclusive ranges or occupy different habitats. For instance, where the ranges of different species of pocket gophers meet, they generally do not overlap. Where the leopard frog occurs alone, it occupies entire ponds; where it occurs with the bullfrog, the latter displaces it to drier peripheral areas.

The most comprehensive observations of this general sort of exclusiveness have been made on birds in New Guinea and nearby islands by Jared Diamond, a soft-spoken ecologist from UCLA who also has had a brilliant career as a physiologist.

Diamond, like MacArthur before him, is a talented field naturalist. He has spent many years doing fieldwork in the Southwest Pacific, carefully mapping the distributions of the many birds of the area. On his numerous trips to New Guinea, Diamond has traveled extensively on foot with local guides, surveying the bird fauna in very remote areas, and recording their songs (he hopes eventually to record songs of all 600 or so species). One of the groups he has studied there is the bowerbirds. Male bowerbirds make not just nests but elaborate structures. One species constructs huts one or two yards in diameter; another builds stick towers several yards high; and still another creates moss platforms with parapets. These bowers are adorned with brightly colored objects: shells, flowers, fruits, and (when they can swipe them from settlements) manufactured items such as bottle caps, tooth-brushes, and coins. High in the Foja Mountains, Diamond even redis-covered a spectacular long-lost species, the yellow-fronted gardener bowerbird.

But important as that discovery was (it helped him work out the origins of the male bowerbirds' strange behavior), it was not as impor-tant as a more general one—that of the "checkerboard" patterns of distribution of a variety of Southwest Pacific bird species. These are patterns in which certain combinations of birds never seem to occur together. For instance, two species of cuckoo doves of the genus *Macro-pygia* are found in the Bismarck Archipelago, off the northeastern coast of New Guinea. Of thirty-three islands sampled, fourteen have one species, six the other, and thirteen have neither. No island has both. Two species of *Pachycephala* flycatchers, five honey eaters, and eleven white-eyes, among other genera, show the same kinds of patterns in New Guinea and nearby islands. Some combinations of species seem to be "forbidden"—they are never found in the same place. Diamond interprets these patterns as indicating competitive exclusion by mem-bers of the same guild. In other words, their niches are too similar—they overlap too completely on one or more axes—for the birds to coexist. The diversity of these bird genera on any given island is limited by the great similarity of the species.

In contrast to these patterns in which members of guilds do not occur together, a guild whose members often did co-occur was carefully studied by Diamond. This guild consisted of eight species of New Guinea fruit pigeons that live in the canopy of the lowland rain forests and dine on soft fruits. At some localities, all eight species of fruit pigeons are present—they can coexist because they partition their re-sources. The fruit pigeons themselves form a graded series in weight from about two to thirty ounces, each roughly one and a half times the weight of the next lighter one. The larger species eat larger fruits than the smaller species; apparently they do not find it profitable to compete with their smaller relatives for little fruits. The smaller species, more-over, forage further out on branches that will not support the weight of the larger birds. Thus the species divide up two resources: food and feeding positions.

Resource partitioning when members of a guild have overlapping distributions is a widespread phenomenon. David Inouye has shown that different species of bumblebees around the Rocky Mountain Bio-

A black rhino wallowing in mud. Note the protruding, prehensile upper lip of this African browser, which eats a variety of shrubby plants. The horns, consisting of compressed fibrous hair, have a high market value because powdered they are thought to have aphrodisiac properties by Indians and Chinese and are valued for making daggers given as gifts in Middle Eastern puberty rites. As a result poaching has decimated populations of this once-common creature.

A white rhino, which unlike the black rhino, is a grazer. Note its straight upper lip, which gives the rhino its name, which comes not from its color but from the Dutch *weit*—wide—describing the shape of its mouth. This rather docile rhino is a somewhat lighter gray than the more dangerous "black," and is even rarer and less widely distributed. The two species, where they occur together, neatly partition their food resources.

logical Laboratory have tongues of different lengths. This makes each bee species more efficient at extracting nectar from a different kind of flower. As a result, different bumblebee species use different resources, as each tends to favor the flowers to which its tongue is best adapted. On the Galapagos Islands, species of Darwin's finches that have very similar diets do not occur together, while those that do occur together have different bill sizes and diets. On a more spectacular scale, lions and leopards, where they occur together, partition their prey by size, with the lions taking mostly larger animals and the leopards taking smaller ones. And black rhinos browse while white rhinos graze, again dividing up the supply of edible plants when their ranges overlap.

In addition to the lack of coexistence in guild members that are very similar, and resource partitioning within guilds where members do coexist, the importance of competition in shaping the evolution of communities is often suggested when two closely related species have partially overlapping geographic ranges. In such circumstances, the differences among guild members tend to be stronger in the zone of overlap than where they occur separately. Certain similar species of Darwin's finches that occur together on some islands in the Galapagos and alone on others provide a classic example. On islands where the two species occur together, the beaks of one species are considerably larger than those of the other. Where the two species occur separately, in contrast, their beaks are the same intermediate size. Similarly, where two Asiatic nuthatches occur together, their bills show an average difference in length of about three-sixteenths of an inch; birds taken from areas where each species occurs without the other have virtually identical average bill lengths. Resource partitioning in the zone of overlap is probably the reason in both cases. The possibility exists, though, that the birds also use beak length as one way of identifying mates of the same species, so selection against the tendency to hybridize may be another reason for beak differentiation where the species co-occur.

Another line of evidence indicating the importance of interspecific competition comes from the consequences of invasions of areas by exotic species. One such invader is the extremely aggressive Argentine ant, which was first accidentally introduced into the United States in 1891 in New Orleans and soon thereafter spread widely in the Southeast and was carried to California. Wherever it has gone, native ant species have suffered population reductions. Unlike the impact of different Darwin's finch species on one another, which is due largely to exploitation competition (simply using the same resources that are in limited supply), the Argentine ant operates by interference competition —by directly reducing the access of other species to resources. Wherever it occurs, especially in towns and cities, the Argentine ant drives out other ant species by besting them in direct combat. In addition to North America, the Argentine ant has invaded and conquered areas of Australia and South Africa also.

Similar situations have occurred underwater. The North American slipper limpet, *Crepidula,* has caused an economically damaging decline in oyster populations in some British estuaries where it was accidentally introduced. Here the competition is simply for space to sit and filter food—plankton and dead organic matter—from the seawater.

Over the past two decades, niche theory has been honed into an interesting, if as yet incomplete, framework within which to study the complexities of community ecology. It has also recently been the subject of an increasing controversy, a controversy that transcends questions related to biological communities and goes to the heart of how ecology, and even science itself, is done. The present dispute is about how important competition really is as an organizing force in community structure. The roots of the controversy can be found in an old tendency of animal ecologists to invoke past or present competition between species as the key to explaining many features of animal communities, even though critical experiments demonstrating competition were lacking.

By contrast, *no* parallel controversy has emerged over the importance of competition (for light, water, nutrients) in shaping plant communities. Since the early part of this century, numerous field and laboratory experiments have been done that involved plant species grown alone and in combination. Adverse effects of shading of some individuals by others (either retarding the growth of or killing the shaded plants) have been repeatedly observed. One common demonstration of such competition is seen in "natural experiments" in which a competitor for sunlight has been removed. If, for example, a forest tree blows over in a storm, creating a new sunlit clearing, a whole array of previously suppressed plant species usually springs up in the opening. Similarly, underground contests for resources between roots have repeatedly been shown to occur by means of careful experimentation. If the roots of forest trees are prevented from extracting water and nutrients from an area of forest floor (by digging a trench around the area and severing the roots that reach into it from nearby trees), a thick undergrowth develops where previously there were few herbs or shrubs.

But thirty years ago, no comparable body of evidence on the importance of competition between animals had been developed. Instead, competition was largely a catchall explanation of community patterns, backed by little evidence from natural situations or experiments and thus usually taken on faith. Very often this meant that other possible explanations for observed patterns were ignored. A number of ecologists objected to accepting such untested explanations, especially when the reasoning behind them contained a large element of circularity. In 1967, Charles Birch and I wrote in a paper in *Nature* that "a simple rewording of the problem is accepted as a solution to the problem. For example, the mutually exclusive ranges of X and Y are 'explained' by 'competition.' How do you know there was competition? Answer, because each 'obviously' excluded the other from its range! A *sine qua non* for rectifying this way of thinking is the more rigorous proposal and testing of falsifiable hypotheses." Such protests created pressure to determine by experiment the importance of competition in nature, and gradually the appropriate studies were carried out.

But in 1977, Dan Simberloff and some of his colleagues renewed the competition debate with a strong criticism of niche theory. They directed their attack at, among other things, Hutchinson's ideas about size differences among competitors and Diamond's checkerboard distributions. The arguments tend to be involved and complex even for ecologists to follow, but one can gain something of their flavor without delving too deeply.

To see what the controversy is all about, let's look at the discussion of Diamond's checkerboard distributions, where, you may recall, there appeared to be "forbidden" combinations of species. Basically, the Simberloff group has asked: How can we know that the checkerboard pattern is not just the result of chance, rather than of competitive exclusions? This is not only a reasonable question, it is a perfectly standard query in science. It is related to the following question asked by someone who has just seen a flipped coin come up heads ten times in a row: "Is this a chance result with an honest coin, or is the coin two-headed?"

The probability that an honest coin will come up heads ten times in a row is easily calculated; it will happen roughly once in every thousand sequences of ten tosses. Thus a reasonable conclusion would be that the coin was two-headed or the flipper very talented—or, to put it another way, something other than chance produced the result (since the odds are less than one in a thousand that it *was* chance). Statistical procedures are commonly employed by ecologists to decide whether a given result is most likely happenstance or the consequence of some interesting biological process (just as statistical procedures can be used to test the "honesty" of a coin).*

In their re-examination of the checkerboard data, Simberloff and his colleagues used a computer to construct many hypothetical patterns of distribution of bird species on the same islands as were studied by Diamond. The computer distributions were designed to emulate those that might occur if only chance and the physical environments (available habitats) of the islands determined which species lived where; there were no competitive exclusion rules. The computer, in essence, created each hypothetical distribution by shuffling the bird species of the fauna and dealing them out to the actual number of islands at random, the only constraint being that each island received the number of species actually found there. That is, an island that hosted twenty species got twenty—but a random assortment rather than the twenty actually found. The distributions Diamond had recorded in nature were then statistically compared with those produced on the computer.

What the Simberloff group found was that checkerboard and other complex distributions described by Diamond were similar to some of the computer-generated distributions. Simberloff and his colleagues therefore concluded that the checkerboard patterns are not evidence that competition is important in determining the structure of the island bird communities. Basically, they claim that those unusual-appearing natural distributions are similar to the sequences of ten heads or tails that show up when enough sequences of ten coin flips are tried. If you look at enough different kinds of birds on enough islands, some will have checkerboard distributions just by chance. The Simberloff school therefore prefers a "non-interactive" view of the distributions. That is, who is where depends primarily on the dispersal ability of each species, the physical conditions in which each can thrive, and chance; but not on the outcome of competition.

The argument does not rest there, however. Both the Simberloff group's use of computer distributions and the ways in which they have been constructed have been strongly challenged by Diamond and his colleagues. The challenges are rather technical and, in my view, correct. One important counterargument is based on the difficulty of defining

---

* A little more detail on the use of probabilities in science in general, and in studying checkerboard distributions in particular, is given in Appendix C.

the appropriate array of bird species to be included in the computer reshuffling. Should it just be, say, members of a single guild, or should it be all the birds of the islands, including shorebirds, hawks, songbirds, etc. No agreed-upon solution is at hand, but the result of including too many miscellaneous species in the array on which the computer distributions are based is clear: the chance that the computer output will include a distribution identical to an actual checkerboard distribution will be greatly increased. Such a checkerboard distribution could thus be spuriously rendered "expected"—that is, caused only by chance—even though it was really caused by competition. The problem is analogous to trying to determine the likelihood of tossing ten heads in a row. If the number of sequences of ten tosses is twenty, the occurrence of ten heads just by chance would be very unlikely; if the number of sequences is twenty thousand, ten heads in a row can be expected to turn up several times.

Another important counterargument is that the computer distributions may have been constructed with the effects of competition already concealed in them. Remember that they are based on the reshuffling of actual distributions. If competition is an important force in structuring communities, it will already have influenced the numbers of bird species and their distributions. Therefore the actual distributions of birds are being compared with computer-generated ones that may in fact include the influences of competition, and not with distributions free of its effects. If no significant difference between actual and "competition-shaped" computer distributions is found, that can then hardly be construed as evidence that competition has not been important in producing the distributions in nature. The jury is thus still out on the checkerboards, but, in my opinion, the verdict will eventually be that competition is guilty of causing most or all of them.

The argument over checkerboard distributions reflects a fundamental difficulty faced by ecologists and other scientists working with extremely complex systems. Many sciences, particularly physics, chemistry, and molecular biology, are largely reductionist—seeking to discover the structure and function of each cog and bolt in nature's machinery. Ecology and evolutionary biology tend to be more historical, holistic, and comparative in their approach. Ecologists and evolutionary biologists seek to understand such things as why organisms performing the same function are often so different, and how and why they developed their present variety. They basically are trying to comprehend the rationale for present structure and operation of the entire apparatus, as well as the history of its construction (which itself might shed light on the rationale).

These two approaches to the world are not mutually exclusive, but most scientists seem to have a natural bent toward either reductionist or holistic investigations. Reductionists have little trouble in finding sensible problems to investigate. They almost always try to answer some version of the question: "How does it work?" Their task is to contrive clever experiments to isolate and understand the function of each part. Answers tend to be either yes or no—either the sequence of nucleotides in DNA determines the sequence of amino-acid residues in a protein or it doesn't. And the reductionists in molecular biology have been increasingly successful at finding answers.

But in holistic sciences, such as ecology, defining the question is often the most difficult part. Suppose, for example, the question is stated as "Does competition occur?" The occurrence of competition has been shown experimentally both in the laboratory and in the field, so the answer is "Yes." But that is not a terribly satisfying answer; many things can and do occur that are of little general importance—triple plays in baseball and TV bloopers, for instance. So that question was not the best.

The sort of question ecologists really want to answer is more like "How important is competition in shaping ecological communities?" Leaving aside the difficulties of defining "important" and "shaping," there are two nasty dilemmas facing an ecologist interested in this question. First, there is no yes-or-no dichotomy; the answer lies somewhere along a continuum from unimportant to all important. The second dilemma is how and where to look for the answer. An analogy may help give you a feel for the second problem.

The question of how important competition is in shaping animal communities is somewhat like the question of how important religion is to Americans. Suppose you were asked to determine the actual importance of religion to the people of the United States. You would first have to find clever ways of discovering the true feelings of individual Americans. This would not be such a simple task, since people often do not give honest answers to pollsters, especially on an issue like this, where there may be social pressure toward appearing religious even if one is not. Having found a way of discovering people's true feelings, you would then have to devise a strategy whereby the sample of people investigated was properly representative of the entire population of more than 240 million. All in all, an apparently simple question turns out to be quite complicated to answer.

Ecologists are continually faced with similar challenges. Going back to the earlier example, to determine the importance of competition, they first have to devise experiments to determine whether competition is operating in a guild of animals, and how effective the competition is in influencing the diversity, form, and behavior of the animals. And then they must sample different kinds of animal communities in some way that gives reasonable confidence that the importance of competition in animal community biology as a whole has been established. That process has been under way for some time and has recently accelerated under the impetus of the challenges of Dan Simberloff and others.

The last two decades have seen an upsurge in the use of field experiments to test for the occurrence and strength of competition in nature. One of the most thorough of these studies was done by my Stanford colleague Jonathan Roughgarden and his students on several islands in the Caribbean. These are great places for doing research—if you do it in the warm waters over the coral reefs that fringe them. It takes real dedication, however, to pursue biology in the hot interiors of the islands. That is where various ecologists, including Jon and his troops, have worked to understand how competition affects the lives and evolution of *Anolis* lizards.

Anoles are members of the large lizard family that includes iguanas and horned lizards. The cheap "chameleons" sold in pet shops are actually the common green anole of the southeastern United States. Anoles are predators, dining largely on insects. Males have brightly colored throat fans that they extend when they are courting a female or displaying to defend their territories against other males. The displays include a rather silly-appearing series of push-ups that nevertheless are taken quite seriously by other anoles.

Each of the Lesser Antilles, the arc of small islands stretching between Puerto Rico and Trinidad, is occupied by either one or two species of *Anolis*. They are extremely abundant—often as many as fifty to seventy-five individuals per one hundred square yards. This density of predators is only possible because the lizards are cold-blooded "ambush" predators (ones that sit still and then lunge at passing prey). They apparently have taken the place of a guild of insect-eating birds that in other areas search out and devour bugs near the ground. In terms of energy consumption, however, Roughgarden estimates that it takes about seventy-five anoles to equal one anole-sized bird (remember the argument using the apparent scarcity of dinosaur predators as an indication that they were warm-blooded); hence the large numbers of *Anolis*.

Roughgarden worked on St. Maarten, where two species occur, *A. gingivinus* and *A. wattsi*. The former is larger, but the two species are more similar in size than any other two *Anolis* species sharing an island in the eastern Caribbean. *Gingivinus* occurs throughout St. Maarten, while *wattsi* is restricted to the central hills. In the hills, *gingivinus* perches higher in the vegetation than it does in the lowlands, and *wattsi* perches below it. In order to elucidate the competitive relationship between the two lizards, Roughgarden and his then graduate student Steve Pacala established four square experimental enclosures twelve yards on a side within the lizards' natural habitat. Fences with smooth plastic overhanging tops kept lizards from climbing in or out, and careful trimming of overhanging branches from bushes kept them from dropping in. Lizards were marked from a distance with squirts from a paint gun so that population estimates could be made (in a way analogous to that used for the checkerspot butterflies).

Shortage of food appears to be the main constraint on the growth of anole populations on St. Maarten. For example, where *gingivinus* occurs without *wattsi,* its abundance is correlated with the number of insects caught in tanglefoot traps (plates coated with flypaper goo) set out in the same sites. The evidence for food limitation was reinforced by experiments in the enclosures: when additional lizards were added to enclosures, growth rates of the individual lizards already there declined; and when all the lizards were removed from an enclosure, the abundance of insects and spiders dramatically increased. Furthermore, others have found that when the food supply of natural populations of Caribbean anoles is increased, the growth rates of individuals also increase.

With food in short supply, the existence of competition seems likely. In fact, competition among Caribbean anoles in general had long been assumed, supported by observations of changes in population density and foraging strategy in one species when a closely similar species was present compared with when it was absent. What Roughgarden and Pacala did was confirm the role of competition in these lizard guilds with a series of definitive experiments.

In one such experiment, the impact of the presence of *wattsi* on *gingivinus* was tested. Sixty individually marked *gingivinus* were placed in each of the four enclosures. Two of the four enclosures were also stocked with one hundred *wattsi* in each. Individual *gingivinus* were removed at monthly intervals, weighed, and returned to their enclosures. After five months, all of the experimental lizards were removed and their stomach contents and reproductive conditions were examined.

Roughgarden and Pacala found that the *gingivinus* that did not share enclosures with *wattsi* had more food in their stomachs, and more females in the single-species enclosures contained eggs than where *gingivinus* were housed with *wattsi*. Not surprisingly, those in single-species cages had also grown more rapidly than the individuals that had been subjected to the presence of *wattsi*.

For those individuals in the two-species enclosures, the stomachs of the larger species, *gingivinus,* contained insects that, on the average, were bigger than those found in *wattsi*. This indicated that there was some partitioning of the prey resource. In addition, *gingivinus* perched higher in the vegetation in enclosures with *wattsi* than when it occurred alone, indicating further partitioning along the space axis. Under the conditions of the experiment, *wattsi* individuals clearly had a negative impact on those of *gingivinus,* apparently by competing with them for food and space. Other evidence shows the relationship to be reciprocal: *gingivinus* has a strong competitive effect on *wattsi*.

These and other experiments provide good evidence that strong competition is occurring today between these two species of anoles. But the question remains, does competition account for the observed differences in the niches and distributions of the lizards? Does competition from *gingivinus* keep *wattsi* out of the lowlands of St. Maarten? Rather than competition, might the important factor be differing temperature requirements of the two species? For example, maybe *gingivinus* needs warm perches, and *wattsi* needs cool perches. Then the absence of *wattsi* from the lowlands would be explained because the lowlands would be too hot. And the observed upward shift in perch height of *gingivinus* in the hills where both species occur would also be explained —it would simply be moving to more exposed, sunnier perches.

Roughgarden and his colleagues examined this microclimatic explanation and ruled it out. They showed with careful measurements that these two species, where they co-occur, do not occupy positions with different microclimates. They also showed that the higher perches are not warmer for the lizards than those closer to the ground. While the higher perches do get more sunlight, thus warming the animals more, they also get more breezes, which compensate by cooling them.

I have given only some highlights of the anole work, but it should give you an inkling of the amount of effort and the degree of care required to demonstrate conclusively that present-day competition is responsible for an observed pattern of distribution. Fortunately, the *Anolis* system on St. Maarten is especially amenable to study. The lizards themselves are abundant, easily observed and marked, and easily confined. Their food resources are known, and techniques exist for measuring both their abundance and the pattern in which those resources are used. And, as Roughgarden has shown in other work, the two lizard species are each other's principal potential competitors. Unfortunately, as we shall see later, not all systems are so amenable to study.

Other careful experiments demonstrating the importance of competition in nature were done by Mary Price (now of the University of California, Riverside) when she was a graduate student at the University of Arizona. Price studied a small guild of seed-eating desert rodents, consisting of kangaroo rats and three different species of pocket mice. She set up special traps that do not harm the animals to determine their relative use of different microhabitats (e.g., a wide-open area or one close to a bush). And, as in the work on anoles carried out by the Roughgarden group, she constructed enclosures to compare single-species and multiple-species use of microhabitats. She also modified microhabitats by cutting off bushes at ground level to see if this changed the local abundance of any of the species in a predictable way.

Price found that the four rodents clearly differed in their choice of microhabitat in which to search for seeds. In the summer, for instance, the kangaroo rat foraged mostly in large open spaces, and one of the pocket mice foraged in small open spaces. The other two species of pocket mouse both preferred to search under large bushes and trees. But how each species foraged depended also on which other species were present. Thus, whenever a second species was added to or removed from an enclosure, the first changed its pattern of microhabitat use. And, when the amount of open microhabitat was increased by the bush cutting, the kangaroo rat, which normally preferred to forage in large open spaces, increased its density (the number of individuals per unit area) there as predicted. Overall, Price's work strongly suggests that competition is a critical influence in maintaining the differences in microhabitat preferences between the species of this guild.

Few systems are as readily manipulated as those of the anoles and the desert rodents. Nonetheless, the number of field experiments testing for the presence of competition has increased greatly in recent years. More than 150 studies were recently tabulated in a literature search by Thomas Schoener, another theoretically inclined ecologist who, like Roughgarden, has done fine fieldwork with lizards. In those 150 experiments, competition was detected in the vast majority of cases. Plants, carnivores, nectar- and seed-feeders, and scavengers more often competed with each other than did herbivores.

These findings suggest that the differing views among ecologists about the frequency and importance of competition may derive in part from the groups of organisms they use in their research. Jared Diamond, Jon Roughgarden, Tom Schoener, and others who have felt competition to be very important have, in general, done their research on vertebrates, often on carnivores. Dan Simberloff and his colleague, Don Strong, Charles Birch and I, and others who have been critical of the emphasis on competition have, like me, done much of their work with insect herbivores—a group that comprises a major chunk of organic diversity and in which competition appears to be relatively unimportant.

In insect herbivores, competition may often occur, but it also may often have very little impact. For example, one careful study was made of a guild of thirteen stem-boring insects—the larvae of beetles, flies, and moths—that attack native prairie plants in Illinois. Interference competition was discovered, but only between the larvae of one beetle and one moth. The latter was almost always injured and eventually killed whenever they came into contact. Although competition was present, it was apparently too restricted and too weak to shape the community significantly. As a result, the guild showed little sign of resource partitioning. So the moth presumably will continue to coexist with the beetle, despite occasional fatal encounters.

But why doesn't competition seem to play a major role in insect herbivore communities? Its relative unimportance appears to be due either to physical factors or to predators keeping populations small enough that they do not significantly reduce the availability of resources, which ordinarily remain superabundant relative to the needs of potential competitors. Only if two organisms are using the same limiting resource do they compete. Remember, for instance, the impact of weather on the size of populations of Edith's checkerspot butterfly. In twenty-five years of monitoring, we have never observed the populations of the Bay checkerspot on Jasper Ridge become so large that the quantity of food plant became limiting, causing intraspecific competition for food among the larvae of that checkerspot. And different food plants and flight seasons effectively prevent interspecific competition between the Bay checkerspot and its closest relative on the ridge, the chalcedon checkerspot.

Ecologists are now trying to integrate competition and other factors into a general theory of community development. Jon Roughgarden has made a start by creating a mathematical theory for the coevolution of competitors, which may explain the biogeography of anoles in the Lesser Antilles. The theory assumes that there is a continuous turnover of anole species on each island. Suppose a large species invades an island already occupied by a single species of an intermediate "solitary size"—an optimal size for exploiting all the available insect resources. Competition is asymmetrical, in favor of the larger invader, since the best size for exploiting the insect fauna by a lone predator is not the best when a larger competitor appears (and smaller invaders don't make it because they are outcompeted by the resident). The larger invader starts evolving toward the smaller optimal size. The original resident evolves an even smaller size in response to its new larger competitor—which is, in a sense, pushing the resident below the optimum. Eventually, the smaller species is totally excluded, and the new invader is alone and at the "solitary" size—until a subsequent larger invader restarts the cycle. Thus competitive coevolution here would be expected to limit diversity, making occupancy of an island by two guild members a temporary situation.

Roughgarden's interpretation of the present *Anolis* situation on St. Maarten is thus that *wattsi* is on its way out, being excluded by its competitor, *gingivinus*. The two are not diverging from some middle ground in response to competition, as would be predicted by classic competition theory. Instead, the bigger species, *gingivinus,* having a larger impact on the smaller *wattsi* than vice versa, is forcing *wattsi* to extinction.

So the crucial question of how competition does or does not influence which organisms live together is being answered—but not by a single, simple answer. In plants, and in most groups of animals (with the important exception of insect herbivores), interspecific competition seems to play a major role in limiting the number of possible coexisting guild members and physically and behaviorally shaping those that do coexist. And mathematical modeling, accompanied by field tests, is at least starting to elucidate the ways competitive coevolution can work.

Competition, however, occurs largely within guilds, which never include organisms from different trophic levels. Community theory has not yet managed to explain very well the vertical structure of communities—the ways in which the guilds at different trophic levels are tied together. There is a great deal of evidence that coevolution occurs between, say, plants and herbivores or predators and prey, but that's about where it ends. This is one reason why some of the patterns seen in natural communities cannot yet be satisfactorily explained. No one really knows, for example, why there are so many more species in the tropics than in the temperate zones. And there is an ongoing debate about the relationship between the species diversity in a community and its stability (the degree to which it can resist change or can restore itself after it is perturbed).

Why should species diversity be high close to the equator? A number of possible reasons have been put forth to explain the species richness of the tropics. One is that the tropics have remained climatically stable for a long period, allowing species there to coevolve and divide up niches more finely. Furthermore, the mild, stable conditions in the tropics make it easier for numerous small populations to exist—they would be less subject to accidental extinction. Unhappily for this explanation, recent evidence indicates that the tropics have not been very climatically stable over the long term. But another possible answer is that because productivity in the tropics is higher (the green plants there carry on an enormous amount of photosynthesis) and predation there is more intense, competition is reduced, and substantial niche overlap can develop. Or, conversely, competition is claimed actually to be more intense in the tropics, but the high productivity permits many narrow-niched species to coexist.

One thing is fairly certain—increased diversity in the tropics results from the availability of a greater diversity of ways of life, not just from a proliferation of species occupying niches similar to those occupied outside of the tropics. Gordon Orians of the University of Washington has pointed out that much of the increased bird diversity in the tropics can be accounted for by the existence there of fruit- and nectar-eaters and of insect-eaters that sit quietly on branches while searching for their prey and then hover while they pick their victims off the nearby foliage.

All of these niches are relatively rare outside of the tropics, but the greater abundance of fruit, nectar, and insects in the tropics makes these ways of life viable, just as the vastly wider variety of plant species creates opportunities for more kinds of herbivorous insects. Diversity clearly breeds diversity, but how and why the process got started remains a mystery. A single acre of tropical forest often contains many dozens of large tree species all doing essentially the same thing—why not just one species expert at taking advantage of the warmth and the abundant sunlight and moisture? The answer may lie in the greater disease and pest vulnerability of monocultures (stands of a single species). But no one is certain.

So much for the state of our knowledge on tropical diversity. What about the relationship of diversity to stability? A number of plausible stories can also be constructed to explain why species-rich communities should be stable. One aggravating problem with them, though, is that even the terms "diversity" and "stability" are subject to different inter-pretations. Should diversity be considered just the number of species in an area, or should their relative abundances be considered? Should the characteristics of the community members be considered? Is a herbivore community consisting of one hundred cows and one hundred horses as diverse as one containing one hundred cows and one hundred grass-hoppers? And is a community stable as long as its species composition does not change? Or should the constancy of population sizes be a factor? Perhaps the ability to return to a previous state following a change (its resilience) should be the defining characteristic of a stable community. On the other hand, maybe it ought to be resistance to change in the first place or the stability of the functioning of the ecosys-tem of which the community is a part (to be taken up in the next chapter).

In spite of these complications, progress is being made toward understanding how diversity and stability are related in natural com-munities. A classic example is a study by Robert Paine of the University of Washington of a community in the rocky intertidal zone of the Pacific Coast. He looked at a community of fifteen species of barnacles, lim-pets, mussels, and other shelled creatures of the rocks, which were all attacked by a single species of starfish. He showed that when that predator was removed, the community was destabilized and its diversity dropped to only eight species. Once the pressure from the predator was removed, competition became the dominant interaction—and some of the former prey species crowded others out. Thus, in this system, predation acted to maintain diversity. The removal of the starfish caused a small reduction in the diversity of the entire community (there was then one less species). But the starfish was a keystone predator in the system, and its absence destabilized the prey community and caused a further collapse of diversity.

Tree species can play an analogous keystone role in tropical forests. Around the research station of the Organization for Tropical Studies at Finca La Selva, Costa Rica, the canopy tree *Casearia corymbosa* is pivotal to maintaining the diversity of the bird community. Although only one fruit-eating bird is a really effective disperser of *Casearia* seeds, twenty-one other species of birds eat its fruits, and several are almost completely dependent on it during one fruit-scarce period of the year. If it were to disappear, not only would some of the bird species vanish, but so possibly would other plants that depend upon the same birds to disperse their seeds. *Casearia* is the classic example of a keystone mutualist—a species on which many others depend for beneficial interactions.

Examples of less obvious forest keystones are herbaceous plants that provide nectar to support the adults of a guild of tiny wasps whose larvae parasitize insect herbivores. The latter have considerable impact, through defoliation, on forest trees. Without the nectar plants (and thus without the wasps), damage to the trees would presumably be more extensive.

The existence of various kinds of keystone species, many of which undoubtedly remain undetected, has led ecologists to conclude that disturbance of ecosystems can produce unpredictable cascading effects. For instance, "high grading"—the practice of removing only economically valuable tree species from a forest—may have a much more deleterious impact on the diversity of organisms in the forest than would appear likely at first glance, if keystone species like the *Casearia* were involved.

Unfortunately, keystone species can only be tentatively identified after extensive investigations. Confirming this tentative identification normally calls for an unacceptable experiment—exterminating the presumed keystone species and seeing if its absence engenders some degree of ecosystem collapse. Indeed, there is a twofold problem in studying the responses of communities and ecosystems to perturbations: the logistic difficulty of creating properly controlled disturbances and their ethical justification. And this problem plagues investigations of the relationship between complexity and stability as well.

Increasing our understanding of how communities are assembled and maintained is thus one of the most difficult and exciting challenges for ecologists. But is there really much significance in community ecology beyond that inherent challenge of deciphering all of nature's mysteries? There is, of course. As you will see in the next chapter, the maintenance of biological communities is critically important to the persistence of our civilization—and comprehending how natural communities are organized and how their diversity is maintained is essential to their preservation.

Communities all over the world are under assault from ever-expanding human populations and activities. One form of assault consists of changing the composition of natural communities through the introduction of exotic organisms. Compared to activities such as paving over biological communities to make suburbs, plowing up marginal land for farms, overgrazing, deforestation, and generating acid rains, such assaults seem relatively insignificant, and yet they have caused havoc in numerous natural communities and created enormous problems for humanity. The introduction of *Opuntia* cacti to Australia was an important historical example, but there have been many others. And these assaults continue—the Medfly, Mexfly, kudzu vine, and walking catfish are all introduced organisms that have recently made the news in North America. Indeed, as people become even more mobile, it is likely that plants, animals, and microorganisms will follow suit.

Furthermore, we now face the prospect of genetically engineered organisms being introduced into natural communities at much increased rates. Past experience indicates that greater care must be taken before introductions are attempted. One of the earliest of engineered organisms was the domestic goat—the "engineering," as in other domesticated organisms, having been accomplished by artificial selection (selective breeding) during the course of thousands of years. The goats' introduction, under the auspices of *Homo sapiens,* was a major factor in the destruction of the Mediterranean basin, making a "goatscape" out of that once wooded and well-watered land.

Fears that the new breed of genetic engineers, gene splicers using recombinant DNA technology, will make superorganisms that could cause catastrophic disruptions are almost certainly exaggerated. Even the creation of a bacterial equivalent of the goat seems unlikely; a universally lethal pathogen seems virtually impossible. Newly minted organisms, if they are to cause serious, widespread problems, would have to be superior competitors and relatively resistant to predation. They would have to become dominant in a world already populated with a vast diversity of organisms that have been thoroughly tested by evolution. Nonetheless, the new breed of genetic engineers could well contribute to relatively localized or temporary ecological disasters.

More understanding of communities is required to permit proper evaluation of the risks from introductions. Ideally, we should, for example, be able to evaluate the "invasion resistance" of a given community. It is already known from both population dynamic and genetic theory and from transplant and competition experiments that efforts to prevent introductions do not necessarily have to be foolproof to be effective. If a few individuals of an exotic species get through a quarantine, they may fail to establish and maintain a population, either because of the standard difficulties faced by small populations or, more important in this context, because of the presence of certain predators or competitors (or even, as you will recall, because the outcome of competition may be determined by the initial proportions of the competitors).

Once a pest (that is, an organism that eats, competes with, or interferes with *Homo sapiens*) does become established, deciding how to combat it with minimum danger to humanity and to beneficial organisms depends on a knowledge of community structure. Even today, all too often the response is just to souse it with an "appropriate" pesticide or otherwise attempt to wipe it out. And, of course, attempts to exterminate pests often bring on the development of resistance, destruction of nontarget populations, and direct hazards for humanity. The ecologically sound approach, in contrast, is known as integrated pest management, or IPM. In IPM, an established pest, whether a malarial organism, a corn earworm, or a starling, is seen as part of an ecological community, and techniques are employed not to exterminate it (which is ordinarily impossible anyway) but to keep its population size below the level at which it can cause significant damage.

Often, satisfactory management can be achieved by altering the community to encourage predators that eat the pest. For example, in Washington state, intensive spraying of apple orchards with a variety of pesticides kept codling moths and some other pests suppressed for many years, with new chemicals being substituted as the pests became resistant. Ultimately, however, one class of pests, spider mites, became extremely resistant to chemical control. In response, an IPM program was instituted that was aimed primarily at protecting another mite that was an important predator of the spider mites. Paradoxically, the program also protected a relatively rare spider-mite pest species, the apple rust mite. This was done because it is only a minor pest and serves as an important alternate food source of the predator when the major pest spider mites are rare. Apple rust mites thus help to maintain the predator populations at a high enough level to prevent outbreaks of the serious pest species. Understanding the community produced a satisfactory control system where broadcast spraying had failed.

But the impact of pests and of human attempts to control them is just one element threatening natural communities today. Potentially, of course, the most destructive act humanity could commit, a large-scale nuclear war initiating a nuclear winter, would alter most or all biological communities beyond recognition. But even without such a catastrophe, devastation of communities because of the activities of *Homo sapiens* is already approaching intolerable levels. As a result, humanity is faced with serious and growing problems of maintaining biological diversity. Although few people realize it, that diversity is vitally needed to satisfy a variety of human needs ranging from those for foods, medicines, industrial products, and esthetic satisfaction to the support of essential services provided by ecological systems.

Conservation biology is the subdiscipline of ecology that attempts to develop sound strategies for coping with the crucial problem of the preservation and restoration of organic diversity. One job of conservation biologists is to attempt to understand the changes that occur in the structure of communities—in, for example, predator-prey and competitor-competitor relationships—as those communities become more and more simplified and fragmented. Identification and preservation (or restoration) of keystone mutualists or keystone predators may be the critical first step in solving many conservation problems.

The intricacies of community structure and function thus have significance far beyond simply providing an exercise ground for satisfying the intellectual curiosity of ecologists.

Having sampled what ecologists know and don't know about how communities are organized, it is time to move up one final step in complexity. Even though many mysteries still surround community organization, when communities are seen as parts of systems that include the physical environment, additional important properties, processes, and potential problems emerge.

CHAPTER 7

LIFE SUPPORT SYSTEMS

Ecosystem Ecology

ALL HUMAN BEINGS and human activities are imbedded in and dependent upon the ecosystems of our planet. Ecosystems *are* the machinery of nature, the machinery that supports our lives. Without the services provided by natural ecosystems, civilization would collapse and human life would not be possible.

An ecosystem consists of the physical environment and all the organisms in a given area, together with the webwork of interactions of those organisms with that physical environment and with each other. If I say it fast, it may sound very smooth and scientific—but just what does it mean? To get some feel for ecosystems, let's start by looking in some detail at a real one—one that has been relatively thoroughly studied and is probably familiar to you from nature films—the Serengeti ecosystem of northern Tanzania and southern Kenya, in East Africa. It is a system of plains, savannas (plains dotted with trees), and open woodlands, where much of the action involves large organisms that are relatively easily observed. It is also an interesting system that illustrates how many of the principles discussed in previous chapters are tied together.

The Serengeti ecosystem is more or less self-contained, a plateau of some 10,000 square miles (about as large as Maryland), bounded on the east by volcanic mountains, on the south and southwest by rocky woodlands and cultivated areas, on the west by Lake Victoria, and on the northwest and north by cultivation and an escarpment. The plateau lies just south of the equator and reaches an elevation of almost 6,000 feet in the east, sloping downward to about 4,000 feet at Lake Victoria in the west. In the east, the mountains that border the Serengeti include the famous Ngorongoro Crater. Near the mountains, the Serengeti plateau consists of extensive grassy plains. Farther west, these slowly give way to low hills covered with a sparse woodland of acacia trees, which make up the western portion of the system.

But the Serengeti ecosystem is best defined not physically, but as the area influenced by large migratory herds of one species of large grazer, the wildebeest (also sometimes called the white-bearded gnu)— a strange-looking antelope with a long mane, a prominent beard, a shoulder hump, and horns like a buffalo's. The system is characterized by the presence of the last great assemblage of ungulate (hoofed mammal) species anywhere on this planet—besides the wildebeest, which are its hallmark, giraffes, African buffalo, Thomson's gazelles, impalas, warthogs (the ugly, feisty, and fascinating pigs of semiarid Africa), and others abound. These animals, some migratory and some not, feed upon the grasses and other plants of the plateau, and in turn are preyed upon by lions, leopards, cheetahs, hyenas, and hunting dogs.

While the Serengeti is a relatively self-contained ecosystem, smaller ecosystems, such as the plains or the acacia woodland, can be defined within it. Indeed, if one wished to understand the ecological processes occurring in just one grove of acacias or in the soil under a square yard of grassland, these too could be studied as ecosystems. On a grander scale, Africa and even the entire biosphere are ecosystems, since they consist of communities of organisms interacting with their physical environments, and there is continuous connection among all of them through the atmospheric parts of water and nutrient cycles. So, as we look in more detail at the Serengeti, keep in mind that it is actually a composite of smaller systems and is itself an element in much larger systems.

As with every ecosystem, the broad constraints within which the Serengeti system operates are set by its physical environment. In its equatorial setting, the major factor in that physical environment, the one limiting plant growth, is not temperature, which is fairly constant; rather it is moisture, which is provided by an uneven and seasonal rainfall. The rain, in turn, is controlled by movements of a meteorological phenomenon called the Intertropical Convergence Zone, which travels back and forth across the equator, lagging behind the sun's movements by some six weeks.

When the Intertropical Convergence Zone moves south in late fall, it brings some moisture from the northeast, which produces a small amount of rain, starting around November. The rain, which ends the dry season and nurtures a new crop of grasses, sometimes lasts only until January, but can continue into March. Then the northward movement of the zone brings moisture-laden winds up from the southeast. These winds, originating over the Indian Ocean, provide heavier rain from March to May, allowing the grasses to mature. July through October is the dry season, during which the grasses die back. The eastern hills produce a rain-shadow effect, so the wet-season rainfall varies from an average of some twenty inches in the southeast to thirty inches in the northwest. Dry-season rainfall shows a similar geographic pattern, varying from four inches in the southeast to twelve inches in the northwest. Superimposed on these seasonal and geographic patterns of rainfall is a longer-term climatic variation whose source is not well understood. For instance, for at least a decade before 1971, the average dry-season rainfall in the northern woodlands of the Serengeti was some four inches less than in the years 1971 to 1976.

Climate largely determines the kind of ecosystem that can occupy an area. The rainfall in the Serengeti region is not sufficient to support a woodland with closely spaced trees, to say nothing of the luxuriant growth of a tropical rain forest. On the other hand, moisture is too abundant to allow true desert plants to dominate.

But, in addition to climate, soils play a critical role in determining what sort of ecosystem occurs in a given place. In the Serengeti, the differing water-holding capacities of different soils provide important spatial variation, but on a smaller scale than do climatic differences. Along ridgetops, the soils can receive no moisture or nutrients from upslope and can store relatively little moisture; in addition, nutrients tend to be leached out of them. Below the ridgetops, the nutrient-poor soils are replaced by a series of intermediate soil types leading down the slopes to nutrient-rich, water-retaining soils that have been deposited by flowing water in the valleys.

A regularly graded series of soils also extends across the Serengeti Plain. This plain was originally formed by airborne deposits of volcanic materials from the eastern range that includes Ngorongoro. The fallout was coarsest near the base of the volcanoes, and there it formed easily leached, porous soils. Further west in the plain and in the northwest, where finer materials were carried by the wind, the smaller particles formed a more densely packed, more absorbent soil. Thus grazing animals can "move down the gradient" either by traveling from a ridgetop to a valley or by migrating north and west—in either case, moving toward better moisture-retaining soils, which, in turn, support nutritious graze for a longer period of time.

So much for the physical part of the Serengeti ecosystem. Now let's look at the food webs that characterize the biological part of the system. Large herbivores are, to say the least, a major element in the Serengeti. The system is very unusual in this respect today, but it would not have been so a couple of million years ago. During the Pleistocene epoch, great herds of hoofed animals roamed much of Earth's land surface. In North America, camels, horses, bison, and mammoths (among others) were hunted by a variety of predators, including saber-toothed cats. Giant deer, bison, woolly rhinos, and mammoths were a prominent part of the European scene, and they provided food for an array of hunters, including lions. In a sense, the Serengeti ecosystem provides not only a fine example of a contemporary ecosystem, but a window on the past— a glimpse of the Pleistocene.

More than a million wildebeests live within the Serengeti system. They undergo extensive and spectacular migrations in order to find water and the highest-quality forage. During the dry season, the wildebeests usually are concentrated in open acacia woodland in the relatively moist northwestern areas. They may have to move as much as fifty miles a day within the woodland areas if food and water are widely separated, since they normally will not go more than five days without drinking.

At the beginning of the wet season, the wildebeests begin to leave the poor, grazed-over pasture of the woodlands. Often they will move toward thunderstorms, which they can detect from fifteen miles away by sound and from up to sixty miles by sighting the anvil-topped storm clouds, to take advantage of the grass that will sprout after the rains. These herbivores move to wherever the producers are producing. As the plains grasses start to form rich lawns, the animals move eastward to feed on them. Then from January to May they continue in a generally clockwise fashion through the central and eastern plains of the Serengeti. At the end of the wet season, as the grass on the plains dries up, the wildebeests return to the moist northwest.

The circular migration route supports much larger herds than could, for example, be maintained if the animals remained in the northwest—since migrating allows them to take advantage of the nutritious but temporary productivity in the drier plains, and simultaneously reduces the grazing pressure in the acacia woodlands.

In their migrations, the wildebeests are generally accompanied by zebras and Thomson's gazelles—small, light-reddish-brown antelopes with a black stripe on the flank just above their white belly. The three species, however, do not all follow precisely the same route and timetable and do not use exactly the same food. Like members of other guilds, the Serengeti migratory grazers (eaters of grass and other herbs, as opposed to browsers, which strip leaves and tender shoots from bushes and trees) partition their resources. Only during the wet season are they all concentrated in one place, on the upper (southwestern) part of the plain's soil gradient, feasting together on freshly sprouted grasses which have a high protein content.

As the wet season draws to a close, however, the zebras are the first to move off down the gradient, eating the comparatively protein-poor stems of the drying grasses. Following the zebras come the wildebeests, which feed more on leaves and the sheaths at the base of the leaves—plant parts containing more protein than the drying grass stems and made more accessible by the zebras' removal of the stems, which formed the upper layer of the herbaceous vegetation.

Then it is the turn of the Thomson's gazelles to occupy the partially mowed "lawn." The "Tommies" feed heavily on protein-rich forbs (nongrass herbs), made accessible by the previous grazers, in addition to the grass sheaths and leaves. Their slender muzzles enable them to feed much more selectively than either the zebras or wildebeests. Tommies generally avoid areas of ungrazed tall grass, perhaps because their small size (about twenty-six inches at the shoulder, as opposed to twice that for wildebeest and zebra) makes it difficult for them to spot predators there. Thus, in addition to making food more available, the larger

The zebra is the first animal of the Serengeti migrants to move down the soil gradient as the dry season begins, feeding preferentially on the protein-poor stems of the grasses as they dry out.

grazers also create a safer habitat for the Tommies by clearing their field of view.

Thus each migrant tends to alter the structure of the Serengeti vegetation in ways that benefit those that follow. But the sequence of utilization raises interesting questions about the physiological ecology of hoofed animals. For example, why can the zebra, a nonruminant, exist on a diet lower in protein and higher in difficult-to-digest cellulose than the wildebeest and gazelle, both members of the ruminant cattle family?

Let me explain what this question means. All large animals (including ourselves) are aided digestively by microbial fermentation in their guts. Beneficial digestive relationships with bacterial and protozoan mutualists reach an especially high degree of development in grazing herbivores, however. Only with the aid of microorganisms living in their digestive tracts can these animals break down the insoluble cellulose in plant cell walls, extract energy from them, and make accessible the protein and other nutrients contained within the cell.

The ruminants (cattle, buffalo, and antelopes) appear to have the most efficient apparatus for using microorganisms to aid in extracting

Wildebeests ("gnus") follow the zebras, feeding more on leaves, which contain more protein than the stems mowed by their predecessors.

Thomson's gazelles ("Tommies") stand only about 2½ feet high at the shoulder. In the Serengeti they graze areas after the zebras and wildebeests have made accessible the small, nongrass herbs that the gazelles can eat selectively because of their slender muzzles.

protein from their food. Their digestive tracts have an enlarged section, the rumen, in which chewed-up plant materials and the microbial mutualists are mixed together and held for up to several days. Ruminants get their name from the habit of "chewing their cuds"—returning partially digested food from the rumen to the mouth for further mastication. Chewing the cud increases the surface area of food exposed to the microorganisms and enzymes in the rumen, and the addition of saliva helps to neutralize acids produced by the digestive process there. By the time that process is complete, the cells are thoroughly demolished and their contents made available for absorption in the intestine. The ruminant's system is efficient but slow.

But given that ruminants are well designed to extract nourishment from tough plant material, how is it that the nonruminant zebra can extract enough protein and food energy from cellulose-rich, low-protein forage that cannot support the ruminant wildebeest? Microbial breakdown of cellulose in horses (the zebra is basically a horse) takes place in the intestine after the digestion of protein in the simple stomach. In the horse's intestine, however, the digestive process is roughly twice as fast as in the rumen of a cow (and presumably of the closely related wildebeest), and food passes through horses at about double the speed it moves through ruminating cattle.

The zebra's secret, therefore, is that it is able to process the low-protein grass stems at a rate sufficiently high to permit it to maintain itself, even though it extracts less protein and energy per pound of food eaten. Thus the speed of a zebra's digestion compensates for its lower efficiency, and the zebra can thrive on a diet of low quality that would starve a wildebeest—as long as the zebra can eat as much of the low-grade vegetation as it wants.

Like the zebras and wildebeests, nonmigratory (resident) ungulates also tend to divide up their environment and the resources of the Serengeti. Two resident antelopes have ranges that broadly overlap that of the resident African buffalo. These are the robust black-limbed topi, which has a shiny, rich red-brown coat, and the sleek impala (males of which have extremely graceful, lyre-shaped horns). But the topi predominate in the short-grass open plains, the buffalo favor tall-grass areas, and the impala prefer acacia woodlands. When all three species feed together, as sometimes occurs, the three show some resource partitioning: the buffalo mows the grasses relatively unselectively; the topi eats leaves from grasses of medium height; and the impala picks out

African buffalo with mutualistic oxpeckers perched on its back. The birds eat ticks and bloodsucking flies, aiding the buffalo while they obtain food for themselves.

An adult male impala. This graceful antelope is a resident of the acacia woodlands of the Serengeti.

green leaves from bushes and short grasses and eats seeds and fruits.

Of course, the Serengeti migratory herbivores also interact with the resident ones. For example, the zebras and wildebeests mow the tall grass as they pass, probably benefiting the impalas, which normally avoid it. But the migrants must also reduce the total grass available to the impalas, causing them to start eating the limited supplies of non-grass leaves, seeds, seedpods, and fruits sooner than otherwise would be necessary.

So the large Serengeti herbivores interact with one another. But what happens to the plants; how do they respond to the attentions of so many plant-eaters? The total impact of the community of herbivores on the Serengeti grassland has been the subject of intensive research. From this work, we know that, on the average, the ungulates consume about 20 percent of the aboveground production (the weight of plant material excluding roots produced per unit area per year) of some two and a half tons per acre of plants in the tall grasslands of the Serengeti—thus these herbivores eat about a half ton of plant material per acre each year. But averages can be misleading (there is an old joke about a statistician who drowned in a lake that averaged only two feet deep). During the wet season, food is superabundant—in April, some twenty-five times the food requirement for that month is produced. But from July through September, production ranges from none (in July) to no more than slightly over half of the grazers' requirements. Thus, at some times of the year, food is in very short supply and pressure on both the animals and the plants becomes severe.

Studies using exclosures (fenced areas from which large herbivores are excluded) have shown that grazing on the Serengeti increases the diversity of grasses while reducing the average height of foliage. Grazing also causes individual plants to produce more shoots (aboveground plant parts) than roots, thus increasing aboveground production. The overall impact of grazing on plant production (as opposed to the amount of that production that is consumed) remains controversial, however.

Most importantly, the grazing of the animals is responsible for the very existence of the grassland. Grasses are generally more grazing-resistant than forbs or the seedlings of shrubs. In the latter two, the special tissues responsible for the growth of shoots are at their tips, where they are vulnerable to removal by grazers. In grasses, by contrast, those tissues are just above each joint of the stem and therefore closer to the ground, where they are much less likely to be eaten. These relatively soft areas are also protected and supported by the sheaths at the bases of the leaves. In addition, the blade of the grass leaf has a growth area at its base so that, unlike the leaves of forbs, it can continue to grow if its tip is eaten. Furthermore, branching in grasses occurs primarily at ground level, and the resulting new growth is often extended and reproduced by out-of-the-way, unlikely-to-be-grazed stems growing horizontally underground or along the ground surface.

If a grass plant is smashed flat, its special growth tissue grows more rapidly on the downward side, with the result that shoots are quickly bent upward once again. This, in combination with the spread by horizontal stems and the ability of leaves whose tips have been grazed off to continue to elongate, makes grasses extremely resistant to trampling and overgrazing by hoofed animals. The overall result of the presence of large herds of hoofed animals in the Serengeti is to alter greatly the competitive relationships of the plants, assuring the continued dominance of grasses over forbs and shrubs and the perpetuation of a grassland ecosystem.

Grazers have other important effects as well. For example, the plant populations growing outside of permanent exclosures on the Serengeti proved to be genetically different from those of the same species inside. Those outside tended to be dwarfed and to grow more closely to the ground as a result of the selection pressures applied by the grazers. Indeed, the Serengeti ecosystem appears to be one of the few remaining great reservoirs of grazing-resistant plant genotypes—a volume potentially of incalculable value to humanity in the "genetic library" of natural ecosystems.

In the food web of the Serengeti ecosystem, the herds of large herbivores support in their turn large predators: lions, leopards, cheetahs, hyenas, and Cape hunting dogs. And, not surprisingly, those carnivores also partition their resources, with each species taking a somewhat different array of prey or hunting in a different way or time of day.

Lions (which weigh 200 to 400 pounds) are found in both the plains and woodlands of the Serengeti. They are organized into "prides," usually consisting of ten to twenty adults and near-adults. The prides are controlled by lionesses, which defend territories against other prides and expel excess subadult males from their own pride. Lions feed primarily on zebras and wildebeests when those migrating herbivores are within their territories. Their prey among the nonmigratory ungulates includes some giraffes and buffalo (which otherwise ordinarily only fall prey to human beings), as well as warthogs and resident antelopes when the migrants are not around.

Lions are ambush predators; they stalk their prey, mostly at night, and attempt to catch them in a short sprint. To be accurate, however, I should say that lionesses stalk their prey, since they do the vast majority of hunting for prides—while the males just lie around, mate, and co-opt a lion's share of the kills. Lest male readers of this book become too envious, I must remind them that lions normally are only temporarily associated with a pride and must be constantly on guard against other intruding males, which might evict them, take over the lionesses, and kill the cubs sired by the dispossessed lions (remember inclusive fitness). Lions, in fact, spend much of their lives in bachelor prides. And they live only about eleven years as opposed to sixteen or so for lionesses —battles over the females take their toll.

The other lionlike cat of the Serengeti, the leopard (75 to 130 pounds), is confined to the woodlands and generally takes smaller prey than the lion. Leopards are solitary, nocturnal stalk-and-sprint predators. Their prey choice overlaps with the lion's at the upper end of the size range (e.g., Thomson's gazelles, topi, and an occasional zebra), but they feed extensively on tiny antelope such as Kirk's dik-dik (fifteen inches high at the shoulder), small carnivores, hares, and birds. Thus lions and leopards partition their prey largely by size.

Cheetahs are as large and heavy as leopards, but much less lionlike and more slender in build than the other big felines. They hunt in the daytime, taking small antelopes and hares which, after stalking, they run down at speeds of up to sixty miles-per-hour. The classic long-distance sprinter, a cheetah may chase its prey at high speed for as much as 350 yards, as opposed to a maximum of 200 yards for a lion. Like leopards, cheetahs tend to be solitary hunters—though they may live as family groups or two males may form a bond—and they and the leopards partially partition their prey resources by hunting at different times, as do the hyenas and hunting dogs.

Hyenas hunt the plains at night and in the early hours of the morning, in groups of one to three when wildebeests and other antelopes are their prey, and in groups of four to twenty for the larger zebra. They are "pursuit predators," harrying their prey for up to two miles. Cape hunting dogs, the other Serengeti pursuit predator, take much the same array of prey as the hyenas. They hunt mostly by day, however, with the entire

A resting cheetah. This lanky, long-distance sprinter is the only large Serengeti predator that does not also function as a decomposer.

pack (two to nineteen animals) participating in chases of a mile or more' in pursuit of a victim selected by the pack leader. Hunters overeat and then regurgitate food back at the den for those unable to hunt—the old, young, sick, and females with young. Both pursuit predators specialize more on the old, young, sick, or wounded in their prey populations than do the sprint predators—even though the latter probably most often kill animals less fit than the average (usually catching the less alert or slightly slower individuals).

The final major element in the Serengeti system is the decomposers, bringing the food chains full circle, supplying the producers with nutrients, and permitting the ecosystem to continue functioning. Many Serengeti decomposer organisms are similar to those of other ecosystems—flies whose maggots help break down cadavers; beetles that feed on hides or dung; bacteria and fungi that do much of the fine work of reducing organic compounds in dead plants and animals and in feces to inorganic compounds.

But the Serengeti system is unique in including a great variety of large animals that help in the process of decomposition by eating dead animals that they did not kill themselves. While the proportion of total decomposition in the ecosystem that they account for is relatively small (they do not eat plants or small animals), they dramatize this crucial function in a way that the less conspicuous microorganisms, worms, and tiny insects cannot.

The large decomposers include lions, leopards, hyenas, wild dogs, and jackals, all of which scavenge and thus act as decomposers at least part of the time. While there is relatively little competition for living prey among these large carnivores, competition for dead prey is another story. Hyenas get about a third of their food that way, lions 10 to 15 percent, leopards 10 to 15 percent, and hunting dogs 3 percent. Lions are the only predators not significantly interfered with by others. But cheetahs, which alone among large Serengeti predators do not add to their diets by scavenging, lose 10 to 12 percent of their prey to hyenas and (occasionally) lions; hunting dogs lose about half their kills to hyenas; and hyenas and leopards are thought to lose 5 percent or more to lions. Interestingly, the three species of jackals in the Serengeti mostly prey on insects and small mammals. Carrion makes up only about 3 percent of their diets.

Of some 40,000 tons of ungulates that die from all causes (predation, disease, starvation, accident, etc.) annually, the large predators devour some 14,000 tons. The remaining 26,000 tons of dead animals contain about 14,000 tons of soft tissues. Of this, an estimated 12,000 tons are eaten by the most abundant of the large decomposers in the system—seven species of vultures and the marabou stork. The remainder is consumed by insects, other invertebrates, fungi, and bacteria.

The carrion-feeding birds, like the herbivores and predators, also partition their resources and habitat. The nearly turkey-size Ruppell's and white-backed vultures have long, sharp bills and barbed tongues, adapted for cutting and gripping the soft muscles and viscera on which they feed. Whitebacks are primarily lowland birds; Ruppell's prefer the hills.

The less common white-headed and lappet-faced (Nubian) vultures are also almost turkey-size. They have deep, strong, hooked bills well adapted for dealing with tough meat, skin, and tendons. The other large vulture in the Serengeti, the bearded vulture (lammergeyer), is rare and restricted to areas near the eastern mountains. Its strong beak is designed for tough tearing, and its tongue for removing the marrow from long bones, which it shatters by dropping them from a height onto rocks.

Egyptian and hooded vultures weigh only one-third to one-half as much as the larger ones. They have long, thin, comparatively weak bills,

Marabou storks, with sacred ibis in the foreground. These large birds are predators and decomposers, but their beaks are not strong enough to rend flesh so they must snatch scraps away from vultures.

which they use to glean scraps of carrion and carnivore dung and to hunt a variety of small vertebrates and invertebrates. The Egyptian vulture is famous for having learned to crack ostrich eggs by throwing rocks at them. It is largely restricted to the eastern plains in the Serengeti; the hooded vulture is common in plains, savanna, and woodlands.

Finally, the marabou stork in the Serengeti is almost exclusively a scavenger—it rarely hunts insects and has never been seen to attack an ostrich egg. Its long, deep, broad bill cannot be used to tear meat from carcasses. Thus it must snatch what it can away from the vultures, or (as it now often does) rummage in garbage dumps or in the offal pits of abattoirs. It shares with the vultures the baldness characteristic of birds whose diets might lead to badly soiled feathers on the head.

Thus we can see the major outlines of the Serengeti ecosystem: a temporally and geographically varying, semiarid physical setting with energy and nutrients entering grasses, forbs, bushes, and trees, passing through an array of herbivores that includes an exceptional abundance and diversity of ungulates, and then passing through a carnivore trophic level that has an equally unusual number of large predators. In turn, the decomposers—including an unusual number of large ones—make their livings from the energy in droppings and the leftovers of carnivores and recycle the nutrients.

Of course, this general outline conceals most details of the form and functioning of the system. Indeed, except for the unusual (for today) size of the large animal populations, the above description could fit most terrestrial ecosystems. But it is the details, some of which we've already seen, that provide insight into the ways the cogs in nature's machinery have been formed and how they now function. To see what I mean about this sort of insight, let's look more closely at one small aspect of the Serengeti ecosystem: how the combined selective pressures of avoiding predators and obtaining food and mates has helped to shape the social systems of the hoofed herbivores. In so doing, we can see how some of the kinds of behavior discussed in Chapter 3 fit into the big picture.

The antelopes of the Serengeti range in size from the 8-pound dik-dik to the 1,500-pound eland. The tiny dik-diks have higher metabolic requirements per unit of body weight than larger species, and as a result they feed very selectively (with the aid of their slender snouts) on such highly nutritious foods as tender young leaves, buds, fruits, and seeds from a variety of species. Both the need to search out special food items and small size would select against group defense in such a small species; a herd of dik-diks, with their tiny horns, would not daunt any predator much bigger than a frog. The dik-diks, which mate for life, therefore live in pairs with their young on permanent territories in woodland offering thick cover. They communicate by scent-marking (depositing feces, urine, or smelly substances from special glands), since visual displays or calls could reveal their positions to predators. The territories presumably are large enough to contain sufficient food items even at the leanest time of year.

In contrast, larger antelopes, such as impalas, cannot be so choosy about their food—their greater food requirements and the size of their mouths (too big to pick out small buds or berries) mandate a lower-quality intake. Impalas also cannot remain hidden; although they do not migrate, they must move about seeking the best graze and browse; groups of up to a hundred or more shift their habitat preference seasonally. About a third of the males in the group hold territories wherever the females are found at a given time; the remainder of the males form a bachelor herd. Each territorial male marks his territory by his presence (visual signal), roaring (auditory), and scent-marking. Females and young wander over areas larger than the individual male territories, passing in and out of them. As we saw in flocking birds, all the members of the herd also benefit from group alertness and communication about predators. If a predator attacks, the entire impala group reacts as a coordinated unit.

Other antelopes, such as waterbuck, kongoni, sedentary populations of wildebeests (found in places such as Ngorongoro Crater, not in the Serengeti), Grant's gazelles, and Thomson's gazelles also have social systems that share many features with the impala system— especially the holding by males of spatially defined territories within the region occupied by the population at a given time, territories that are held for a considerable period but eventually abandoned when the herd moves on.

Migratory wildebeests, on the other hand, do not have spatially defined, semipermanent territories. Their nutritional needs keep them on the move, so they cannot attach themselves easily to fixed places. Instead they set up tiny, temporary, or sometimes even mobile territories (ones whose boundaries are constantly changing) wherever the females are—defending those territories during the short rutting period. Thus the pattern of males attempting to establish a territory and hence gain exclusive sexual access to a group of females persists in spite of the need for constant movement to find food.

Among buffalo and perhaps eland (researchers are not sure of the latter)—the largest of the antelopes—a different system prevails. Both form large herds that move a great deal. Rather than defending territories, the males simply exercise individual dominance on the spot over subordinate males to keep them away from a female in estrus. There is no defense of a space, only of a female, wherever she might be. Both the buffalo and the eland males solve the problem of access to females by simply staying with them most of the time. Neither of these species has a short rutting season (although the estrus period of an individual is not long), so maintaining a brief defense of a group of females would be to no avail.

All members of buffalo and eland herds cooperate in the defense of their young. This is likely due to the relative inability of buffalo calves to fend for themselves compared to wildebeest calves. Buffalo males presumably do not attempt to appropriate groups of females by cutting them out of the herd because this might expose the calves to predation. The formation of cohesive herds by buffalo and other large species is facilitated by their ability to tolerate relatively low-quality food. Their large size means that they require less energy per unit body weight, and they can thrive by mowing down abundant, evenly dispersed, relatively protein-poor food, such as a field of grass.

We have sampled ways that selection shapes functioning of the various organisms in the Serengeti ecosystem; now we'll consider a higher level of integration: the stability of the system as a whole. In a biosphere increasingly subject to human disturbance, the stability of ecosystems has become a subject of major concern to ecologists, since ecosystems provide a wide array of essential services to humanity. The difficulties of evaluating the stability of communities were touched on in the last chapter. Those problems are part of the difficulty of judging the stability of the more inclusive ecosystems of which communities are but the living parts. Since the Serengeti ecosystem has been so well studied, what can be said about its stability? Can it be disturbed easily? If it is disturbed, does it return to its previous state?

Two large-scale perturbations of the Serengeti ecosystem have been observed. The first was a great epidemic of rinderpest, a viral disease native to the steppes of Asia that attacks ruminants. The disease has apparently been introduced repeatedly into Africa. The epidemic in question may have had its origin in viruses imported with cattle brought by the British from Russia in 1884 during their incompetent (and un-successful) attempt to relieve General "Chinese" Gordon at Khartoum, in the Sudan. The disease first became epidemic in the Horn of Africa in 1889, however, and so the virus may have instead entered Africa in infected zebu cattle brought to Abyssinia (Ethiopia) from India to feed Italian troops.

Whatever its origin, the disease raced southward, reaching South Africa in 1896. In the Serengeti region, it decimated the cattle of the Masai and other herding tribes. This led to horrible famines between 1890 and 1920, with at least two-thirds of the Masai dying. An old Masai eyewitness stated that the corpses of people and cattle "were so many and so close together that the vultures had forgotten how to fly."

By 1890 the disease had infected native ruminants, and the buffalo, wildebeests, and giraffes were disappearing. Among other things,
this led to starvation of their natural predators, and as a result a small
number of lions switched to eating people. By 1898 the man-eating
lions of Tsavo had become famous; in 1920, in an outbreak of man-
eating in Uganda, one lion was reported to have dined on eighty-four
people. The appearance of the man-eaters caused farmers to abandon
their land, and this, combined with the disappearance of the herdsmen
and the native grazers and browsers, led to a reinvasion of plains areas
by brush and woodland. By 1910 or so, the wild ruminants had acquired some resistance to the disease, and their numbers started to
increase again. These herds, in turn, supplied blood meals for tsetse
flies, which further expanded their range into the new brushy areas.
Sleeping sickness carried by the tsetse flies suppressed the human populations even further.

In the 1930s, the tide began to turn. Brush-control programs gradually regained ground from the tsetse-fly invasion, and vaccination
began to reduce the impact of rinderpest on cattle populations. Slowly
the rinderpest began to fade from the native ruminants, first producing
mortality only in yearlings that had not yet acquired immunity, and
finally disappearing in the early 1960s.

The results of the ebbing of the rinderpest epidemic were dramatic.
The survival of wildebeest yearlings increased from 25 to 50 percent,
and their Serengeti population shot up from a quarter-million individuals in 1961 to half a million in 1967. Buffalo increased from 30,000
to 50,000 in the same period. The nonruminant zebras, immune to
rinderpest from the very start, showed no such population change. The
system was returning toward its pre-rinderpest state, although helped
along by human interventions such as vaccination and brush clearing.
(Of course, human transport of cattle had caused the perturbation to
begin with.) The only conclusion that seems reasonable to draw from
the epidemic is that a tiny virus is capable of significantly, and perhaps
permanently, altering a major ecosystem.

On a visit to the Serengeti in 1984, I was told that rinderpest had
reappeared among the buffalo, causing heavy mortality in the Ngorongoro Crater population. The significance of this is not known, unfortunately, for research on the Serengeti ecosystem has declined as a result
of Tanzania's worsening economic condition.

The second major disturbance of the Serengeti was an increase in
dry-season precipitation in 1971–1976. This raised the productivity of

the grasslands at a time when it is usually at a low point. That, in turn, led to a reduction in mortality and thus to further increases in wildebeest and buffalo populations, the wildebeest reaching about 1.3 million by 1977. There was no such increase in other grazers such as the zebras —for reasons that are not entirely clear.

The increase in wildebeests may have had a positive effect on the population size of Grant's gazelles by altering the competitive relationships among the plants. Herbs did better than usual because of the intense grazing pressure on the grasses, and the gazelles eat the herbs by preference. This, in turn, may have been responsible for an increase in the population of cheetahs, which feed heavily on the gazelles.

The larger number of wildebeests might also be expected to benefit their major predators, lions and hyenas. But that was not the case. Nor, for that matter, was the postrinderpest wildebeest population explosion accompanied by irruptions of lions and hyenas. The reason apparently is that the territorial populations of these predators are controlled by food supply (mostly nonmigratory ungulates) during the scarce period when the migratory wildebeests are absent. The predators did, however, respond to the rainfall-induced increases in resident hoofed mammals (topi, kongoni, warthog). In the early 1970s, lions almost doubled in numbers, and the hyena population increased by almost 50 percent. These increases were bad for the hunting dogs, because they suffered both increased interference competition and predation on their young by hyenas. The size of the hunting-dog population declined.

Other elements of the system changed as well. The increased rainfall reduced the frequency of dry-season fires; that in turn fostered the survival of small acacia trees, leading to increased numbers of giraffes. By feeding on the tops of small acacia trees, giraffes tend to prevent the small trees from growing into the mature class. The mature acacias, on the other hand, are often killed by elephants, which push them over and eat the leaves and small branches. A change in rainfall thus could have affected a complex dynamic fire-giraffe-elephant-tree system that ordinarily may change only very slowly because of the long generation times of the organisms involved.

Thus we can see that, in the relatively brief period in which it has been studied, the Serengeti ecosystem appears to have been in flux, responding to both a human-induced virus epidemic and natural changes in climate. There is certainly no sign that the system has some natural equilibrium state to which it tends to return automatically after disturbance, but neither is there any indication that major irreversible change is easily and quickly entrained. If the disease or climatic picture is altered, the system changes in response; if the original alteration is reversed in a relatively short time, the ecosystem may well also tend to revert to its previous state. The paleontological record indicates that the major features of the Serengeti system—including the semiarid climate —have been relatively stable for at least a million years, and possibly for much longer. If there were a permanent increase in rainfall to the point where the major migrants became sedentary, however, the entire structure of the system would change substantially.

But how does the special case of the Serengeti relate to the general properties of ecosystems? All significant ecosystems, of course, are driven by the energy of the sun. That energy evaporates the oceanic waters that eventually rain on the Serengeti and all other terrestrial ecosystems. And, you will recall, the sun's energy is captured by the green plants in the process of photosynthesis. In that process, the energy is converted from the radiant form of sunlight into chemical energy in the bonds of glucose and other carbohydrate molecules. That chemical-bond energy is then used by the plants, and by successive trophic levels up the food chains: herbivores, predators, and decomposers.

Knowing the role of energy is crucial to understanding ecosystems, so it is with energy that we must begin looking for general ecosystem properties. Key aspects of the behavior of energy in all situations, including in the running of ecosystems, are described by the famous first and second laws of thermodynamics. The first law (law of conservation of energy) simply states that energy can be neither created nor destroyed, although its form may be changed (as from radiant energy in sunlight to bond energy in molecules produced during photosynthesis). The second law states that whenever energy—which can be described as stored work—is actually used to do work, some of it becomes unavailable to do further work. Some of it, in practical terms, is lost. Organisms at each trophic level do work in the course of maintaining their structure and metabolism, growing, and reproducing. The energy so used is subject to the inexorable tax of the second law, and the portion taxed away is not available to the next trophic level. The significance of the second law here is that in any ecosystem the amount of energy available to each successive trophic level declines. Thus more energy is available to support plants than herbivores, more to support herbivores than carnivores, and so on.

Two consequences of the second law in the Serengeti are that the biomass (total living weight) of grasses and other grazed plants is much greater than that of the grazers, and that the biomass of grazers is much greater than that of the predators that attack them. There can be huge herds of wildebeests but no huge herds of lions, thanks to the behavior of energy described by the second law. The second law also explains why populations of insect pests are normally much larger than those of the predators that feed upon them: indeed, it explains why one relatively rarely encounters large predators in nature—why there are so many more sparrows and finches than hawks and eagles.

Ecologists usually consider, as a rule of thumb, that about 10 percent of the energy that flows into one trophic level is available to the next. Thus, if green plants in an area manage to capture 10,000 units of energy from the sun, only about 1,000 units will be available to support herbivores, and only about 100 to support carnivores. This means, for instance, that roughly ten times as many people can survive eating a corn crop as can survive eating cattle that have been fed on the corn. The higher on a food chain a human population dines, the less food will be available to it—everything else being equal. But everything else is rarely equal. The logistic problems of harvesting plankton, for example, generally mean that you can get more to eat by fishing than by catching and eating the plankton, even though fish live higher on the food chain.

A decline in biomass with each increase in trophic level is the rule in terrestrial ecosystems, but the second law does not *always* result in a decline in actual biomass at higher trophic levels. That is because the law specifies what happens to *energy flows,* not weight of materials. For example, there can be a greater weight of consumers than of producers if the turnover in the producer community is significantly higher than in the consumers (that is, if the birth and death rates of the producers are much higher than those of the consumers). In the English Channel, for example, a rapidly reproducing community of tiny phytoplankton supports a greater weight of more slowly reproducing zooplankton, fishes, and other consumers. But the second law is not violated since, over time, substantially more energy flows through the producer trophic level than flows through the consumer level.

This point tends to be a little confusing, so consider a partial analogy. Suppose your pantry were magically restocked every night with three pounds of food, which happened to be exactly the amount you eat every day. Your biomass is many times higher than the biomass of food in the pantry, but the rate of turnover is much higher in the pantry (since the pantry food is "born" and "dies" each day).

The second law also explains a major difference between energy and nutrient flows in ecosystems. Because of the steady loss of energy as it is transferred up food chains, it is evident that energy must make a one-way trip through ecosystems. Energy, as the second law states, cannot be recycled; it can be used only once. The portion that is used at one trophic level is not available to the next. In contrast, nutrients used by an acacia tree can also be used by a giraffe that eats the tree, used again by a lion that eats the giraffe, used yet again by a vulture that eventually consumes the lion's body, and, when the vulture sits in the acacia tree and fertilizes the soil at its base with droppings, reused by the tree. Nutrients can be recycled—indeed, they do move in circular paths through ecosystems. That is why ecologists speak of nutrient cycles, but not of an energy cycle.

The acacia tree, like the other plants of the Serengeti, with help from hardworking microorganisms, takes up from the soil the mineral nutrients—nitrogen, phosphorus, potassium, iron, calcium, and so forth —necessary for life. At each transfer up the food chain, the complex molecules containing the nutrient elements usually are broken down into simpler compounds, then the nutrients are reincorporated into the living matter of the next member of the chain. In all food chains, decomposers break down the complex organic molecules in tissues and waste products, transferring the elements from the biological back to the physical part of the ecosystem. For example, certain bacteria make their living by breaking down complex compounds containing nitrogen, thus releasing simpler chemicals that plants can reabsorb and even elemental nitrogen itself (which re-enters the atmospheric nitrogen pool). Their activity is balanced by a different group of organisms that can capture atmospheric nitrogen and convert it to a form usable by plants (and eventually by consumers). Still another group of bacteria gets its energy by decomposing complicated organic compounds containing phosphorus and producing the simple, inorganic phosphate that plants can use. In such ways, the nutrient cycles continually lubricate the machinery of nature.

Thus all ecosystems are driven by energy, and they cycle nutrients, largely through the agency of organisms adapted both to the physical conditions of the system and to each other. If ecosystems were cut off

Desertification in the Sahel. These tribesmen in Upper Volta in 1973
have watched their herds of cattle vanish from starvation as their
pastures gradually turned into wasteland.

from their energy source (as in a nuclear winter), they would cease to
exist. If their nutrient cycles are interrupted, ecosystems collapse. If
those cycles are substantially altered (as in climatic change), the character of the system will profoundly change.

Homo sapiens, of course, can dramatically alter ecosystems with
tragic results. Nowhere is this now more evident than in the famine-
plagued semiarid areas of Africa. There, as in many other areas of the
world, human activities are increasing the areas of desert at the expense
of moister and (from a human standpoint) more desirable ecosystems.
Thus ecosystems that can sustain rather large numbers of *Homo sapiens*
are being replaced with others that cannot. For example, the southern
edge of the vast Sahara desert has been moving inexorably southward.
In the Sahel, the territory just south of the Sahara, and in Africa in
general, cattle are playing a major role in this desertification.

The reason is straightforward. Much of the continent is too arid for
plant cultivation without irrigation, which is, for most African societies,

too expensive. (Irrigation is also temporary—since dams and canals eventually silt up, underground water supplies can be drained, and irrigated land is often ruined by salting up and waterlogging.) But human beings can support themselves, in the absence of irrigation, by using cattle as intermediates between themselves and the naturally oc-curring plants, which can use the little water available and which the cattle can eat but people cannot (the fermenting apparatus of our guts is comparatively puny). In Africa, cattle are more than suppliers of food and hides, however. They are also viewed as a direct indicator of wealth by many peoples, such as the Masai.

From an ecological point of view, though, domestic cattle are a source of environmental deterioration in hot, semiarid climates. Cattle (and goats and sheep) must walk daily to a water supply to drink. This consumes a good deal of energy and slows the rate at which they gain weight. It also results in the trampling of valuable grasses and compact-ing of the soil surface, especially around water holes—where bare, pounded areas continually increase in size. On the other hand, wild herbivores have much less need to drink. Some, such as eland, oryx, and Grant's gazelle probably do not have to drink at all, obtaining all the water they require from the vegetation they eat. Others such as impala do drink some water, but still require much less than cattle.

Most native African herbivores also conserve water much more efficiently than cattle. For example, nearly all of the moisture is ex-tracted from the intestinal contents of gazelles before their dry feces are released. Cow pats, in contrast, are produced moist, and they quickly lose ammonia (a gas containing the vital nutrient nitrogen) to the at-mosphere. Cow pats also dry rapidly in the sun and heat up, killing the bacteria and fungi that might speed their decomposition. The flat, dried cow pat also kills the grass beneath it. On the other hand, the dry fecal pellets of antelopes are roughly spherical. They fall between the grass blades, do not heat up, and retain their nitrogen. Rather than tending to create a "fecal pavement" as cattle droppings do, the pellets are readily broken down by decomposers, returning the nutrients to the soil.

Moreover, like most herbivores, cattle have quite specific food pref-erences—they graze some grass species heavily and others not at all. In cattle-raising areas, the mix of species that make up the forage changes, with the kinds of grass not eaten by cattle becoming increasingly com-mon. As we have seen, the native herbivores partition the available plant resources; their diets, to one degree or another, complement each other,

Giraffes browsing on the tops of acacia trees on the slopes of Ngoron-goro Crater above the Serengeti Plain. The giraffes are able to consume plant parts inaccessible to other large browsers, playing their role in the partitioning of resources by African ungulates.

so they can utilize many more types of vegetation than can cattle. For example, the tops of acacia trees eaten by giraffes are a food source out of reach of other herbivores and totally unexploited by cattle.

Not only are the water-conserving native herbivores thus better adapted to the semiarid habitats of the African savannas, they usually do not degrade it physically or chemically, or reduce its plant diversity. Cattle, however, parading back and forth to water holes, leaving destructive droppings, and exercising their food preferences, have been a major engine of desertification on the continent.

All of these differences led wildlife biologist David Hopcraft to conclude that the soundest way to exploit many African grassland ecosystems was not by grazing cattle but by organizing ranches to raise and harvest native herbivores. The idea is not unique, but the experiment that Hopcraft and his wife, Carol, are running appears to be. On their 20,000-acre ranch on the Athi Kapiti plains near Nairobi, they have been transforming ecological theory into practice since 1978.

The ranch is stocked with a variety of grazers and browsers, including antelopes, zebras, giraffes, and ostriches. Cattle are being phased out and may one day be replaced by a native bovine, the African buffalo. For the present, however, the cattle serve as a valuable "control" for comparing costs and meat yields on the same land with those of the native animals. A principal tool of the Hopcrafts' operation is a perimeter fence thirty miles long, specially designed not to injure animals that run into it. A great deal of research is being carried out at the ranch. The dynamics of the various populations are carefully tracked, the food preferences of the different animals are recorded, and veterinarians at the ranch are studying the parasites of the harvested animals.

So far the results of the experiment are exceeding the Hopcrafts' early hopes and expectations. The condition of the range has been *improving*—even though the combined biomass of cattle and native herbivores has increased by some 35 percent in the past few years. Harvesting of antelopes is efficient and more humane than in an abattoir. One night each week, men in Land Rovers spotlight surplus male animals and dispatch them instantly with a high-velocity bullet to the brain. The other animals are not seriously disturbed. Each carcass is quickly processed under the scrutiny of a government inspector.

Could game ranching provide part of the solution to Africa's problems of desertification? The answer will depend on many things. The ultimate yield from the ranch will be determined, in part, by the mix of animals finally established. If the emphasis is put on harvesting smaller animals, such as Thomson's gazelles, rather than larger ones, such as eland, production may be lower because the smaller animals have higher metabolic rates and thus require more forage per pound of meat produced. The optimal mix of herbivores will depend on their exact food preferences and the precise mix of edible plants available—which can probably only be determined by experimentation with different combinations. There is even reason to believe that it might often be advantageous to keep some cattle in the mix, since they are easily herded and can be moved to where they can most profitably graze without doing damage.

But if game ranching is to help roll back the deserts, it will have to produce a product that will be valued by consumers. The Hopcrafts are experimenting with various ways of marketing the lean and tasty antelope meat—including the production of an addicting (to this American at least) jerky—and seem to have established the basic economic feasibility of African game ranching. Costs are substantially lower than

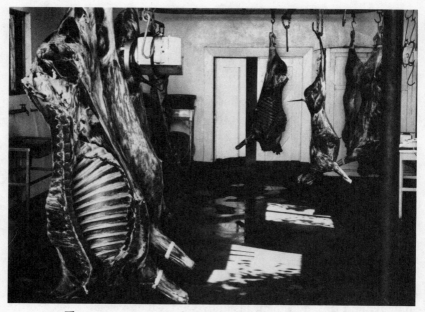

The meat-processing facility on the Hopcraft game ranch. In the foreground are large wildebeest carcasses, in the center distance is a kongoni, and in the right foreground a Thomson's gazelle.

those of raising cattle in that region. Much less water has to be supplied to the animals, meaning less capital must be sunk into wells, dams, piping, etc. And, unlike cattle, game animals do not need to be frequently dipped or inoculated against parasites and diseases. There is also no need for herding and corralling—the native herbivores handle their own predator protection. Indeed, the Hopcrafts' herds have been expanding in spite of almost no predator control. Lions, cheetahs, hyenas, and jackals all inhabit the ranch, and the very small share of the game they get causes little concern.

An additional advantage of the game ranch is the potential for selling the hides. Those from the game should have a much higher value than cowhides, but at the moment the sale of game hides is prohibited by the Kenyan government—quite properly, since most of them are obtained by poaching. Possibly a licensing system can be devised to allow game ranchers to market their hides, like that for mink ranchers in the United States. Combining higher meat yields with the lower costs of production and possible profits from hides, one can foresee a bright economic future for game ranching. From the data gathered so far, it appears that, near Nairobi, the annual yield of lean meat from game ranches could be at least twice the poundage per acre as is taken from the best cattle ranch in the area.

There are, of course, many problems still to overcome. The main ones involve breaking traditions: traditions of what meat is good to eat; traditions among scientists in animal husbandry for whom the idea of game ranching is too novel; traditions among African pastoral people such as the Masai, for whom—as noted above—cattle are the main symbol of wealth and cattle raising is a way of life central to the culture.

In Africa, however, the benefits of breaking with tradition would be enormous. In the face of extremely rapid human population growth, game animals are fast disappearing from the continent; even supposedly inviolate national parks are under intense pressure from expanding agriculture and poaching. Food supplies per person in most African countries south of the Sahara desert have declined by well over 10 percent since 1970, as desertification proceeded and population growth outstripped the gains in food production. So, even if game ranching were slightly *less* profitable than cattle ranching, game ranching would be well worth subsidizing. The long-term costs of continued desertification are incalculably higher than any conceivable subsidy that would be required to make game ranching an attractive enterprise.

But the subsidies have been working against game ranching. Desertification caused by overgrazing of semiarid lands by traditional domestic animals has been encouraged by shortsighted "aid" policies aimed at converting even more land, both arid and moist, to cattle range by exterminating the tsetse flies that carry a sleeping sickness (to which native animals are immune but cattle are susceptible). Recently, DDT spraying sponsored by the European Economic Community was threatening to destroy (by removing the flies so that cattle could be grazed) one of the last strongholds of game in southern Africa, the Okovango swamps of northern Botswana. Game ranching clearly could help to preserve Africa's unique large animals while contributing substantially to the food supply of its human population, but if action is not taken soon to preserve the essential natural ecosystems that can supply reservoirs of game, its time will have come and gone.

Don't think that the idea of game ranching is a benefit that applying ecological knowledge can bring only to such faraway places as Africa, however. Many of the semiarid areas of the western United States are also threatened by desertification, caused, once again, largely by over-grazing of domestic animals, primarily cattle. As a result of this unsound use of rangeland, streams that are not drying up are being polluted, water tables are dropping, land is eroding, and wildlife habitat is being destroyed. Abusive grazing practices on federal land are acquiesced to by the Forest Service and Bureau of Land Management because of the political clout of certain western ranching interests that have grown fat on gigantic government subsidies. The needs of huge numbers of hunters, fishermen, campers, farmers, municipalities, and nature lovers for well-watered ecosystems have been subordinated to the greed of a few who are creating deserts for short-term profits.

The situation is especially pathetic because all grazing on federal land in the western states accounts for only a small percentage of U.S. beef production. The time is long overdue for ecologically sound management of public lands in the American West—including both reductions in cattle and sheep grazing to well within the long-term carrying capacity of the range and the reservation of many areas for the exclusive use of wildlife. Some of the latter could be exploited by game ranching. North American animals eminently suitable for such ranching include deer, elk, moose, bison, pronghorn, and mountain sheep. In Africa, extreme poverty and lack of trained biologists are major barriers to the establishment of sensible grazing systems. Americans do not have these excuses; in fact, some of our best young ecologists are unable to find employment.

Desertification is an example of an ecosystem alteration on a gigantic scale—one kind of system being replaced by another. Changes of such magnitude most frequently occur because of climatic change or because of the intervention of humanity—and both often are involved in desertification. But progressive changes also can be seen within a single ecosystem: the invasion of the Serengeti by brush when cattle ranching was reduced; the appearance of weeds in a newly bulldozed road cut in California; the movement of trees into abandoned farm fields in Vermont; or the settling of corals on the remains of a Japanese Zero fighter resting on the bottom of Rabaul Harbor—all are collectively known as succession.

Trying to understand how and why succession occurs and learning to predict its course have been major concerns of ecologists since early in this century. At that time, the dominant view of succession was based on Frederick Clements' idea that biological communities were "superorganisms." In 1916 he wrote of ecosystems (which he called "vegetational formations"): "like an organism the formation arises, grows, matures, and dies. . . . The life history of a formation is a complex but definite process, comparable in its chief features with the life-history of an individual plant."

Nothing vague here—Clements thought that succession was the same sort of process of development that a chestnut undergoes as it metamorphoses into something the village smithy can sit under. Just as there is an orderly sequence of cell divisions, tissue formation, and production of organs in the development of an individual from a fertilized egg into an adult plant or animal, so he thought that there is an orderly sequence of occupants of a given physical site that leads to the development of a mature climax community. Generally, Clements believed succession was a sequence of large, stationary, space-occupying organisms—plants on the land and invertebrate animals such as mussels, barnacles, and sea anemones on the sea floor. Ecologists of Clements' day and today have recognized that such organisms were key to the kind of total community that occupied any spot—they provide the structure that largely governs the availability of niches for other organisms.

While it has long been clear that the community-superorganism analogy was, to say the least, overdrawn, notions based on it are still important in the study of succession. One is that the early occupants of a site facilitate the invasion of that site by later successional species, and thus the pioneers hasten their own demise. This indeed does seem to occur, especially in what is known as primary succession. (Primary succession occurs after a bare area is exposed to colonization—as when rock or mud is exposed by a retreating glacier or a landslide, a new sandbar is deposited by a shift in ocean currents, or a new island is created by volcanic emergence.)

When glaciers retreat at Glacier Bay, Alaska, pioneer plants—especially prostrate willows—form a mat that traps rock particles and organic debris and starts producing deep, fertile soil. That soil is required by later species, such as alders, and even later the Sitka spruces, which eventually overtop and "shade out" their predecessors, preventing the smaller plants from obtaining enough light for photosynthesis. Similarly, along the shores of Lake Michigan, one species of grass plays a crucial role in stabilizing drifting sand dunes so they can be colonized by other plants and eventually form the substrate on which woodland grows.

But primary succession is much less important in today's world than secondary succession. Secondary succession occurs when an already-occupied area is disturbed, but the soil is not destroyed. Secondary succession naturally follows in the paths of prairie fires and forest-toppling hurricanes. It has begun to repair the devastation of the eruption of Mount St. Helens. But the most common examples today occur because of the activities of *Homo sapiens,* a species whose plows outdo prairie fires and whose chain saws are much more destructive of forests than are violent storms.

Even nonindustrialized people can be potent generators of secondary succession. For instance, in tropical moist forests, people have long practiced what is variously called shifting cultivation or slash-and-burn, swidden, or milpa agriculture. A farmer cuts a clearing in the forest during the driest season of the year. After the felled trees have dried out, they are burned. Many potential pests are killed off by the hot fire, and desired crops are planted in the nutrient-rich ashes. In subsequent seasons, the crop residues are burned instead—but the fire is less hot and the resultant ash less nutritious, and the pest control is therefore less successful. Furthermore, nutrients are progressively lost from the clearing by leaching and by being removed with harvested crops. After two or three seasons, the soil is exhausted, and the farmer moves and starts the process again at another site. After the clearing is abandoned, various grasses and other herbaceous plants gradually invade, followed by shrubs and saplings. As the saplings mature into trees, the dense leafy canopy characteristic of tropical moist forests is re-established. Many of the pioneer occupants of the clearing die in the shade. After decades, secondary succession has restored the original clearing to tropical forest.

People causing secondary succession could be considered part of a natural evolutionary progression, though. Clues provided by our nearest living relatives, gorillas and chimpanzees, suggest that, at one important stage in human history, our ancestors may have been creatures of secondary successional habitats. Suitable foods, especially diverse plant foods, are much more abundant and accessible in regenerating forest clearings and edges than under the closed canopy of a mature tropical forest. So we now may create for ourselves the kinds of habitat in which our ancestors thrived.

But, since the agricultural revolution some 10,000 years ago, humanity also can be thought of as having been constantly battling against secondary succession. Farming, whether in a milpa plot or on a Midwestern factory farm, involves halting the successional process at a stage in which the productivity of the desired plants is maximized. For example, unless it is continuously plowed, planted, and weeded, a farm field in most parts of the eastern United States will gradually revert first to brush and then to a deciduous forest.

The extent of human perturbation of ecosystems is now so vast that understanding the systems' responses—and that means, among other things, understanding succession—has become crucial to human health and well-being. I mentioned earlier that natural ecosystems supply humanity with a series of vital (but largely unappreciated) public services whose curtailment would threaten civilization. Let's look at how these services are supplied and then return to succession to see how it can affect their delivery.

The atmosphere, you will recall, was not always as it is now. One of the services that ecosystems provide is control of its quality. The oxygen that animals depend upon is produced by plants, for example. In addition, a great variety of microorganisms control the concentration of the other major gaseous constituent of the atmosphere, nitrogen (including amounts of nitrogen compounds that play an important climatic role). The nitrogen that makes up almost 80 percent of the air we breathe is the major storage pool in the complex cycle of that element through ecosystems.

Some microorganisms, principally cyanobacteria (blue-green algae) and other bacteria, "fix" nitrogen—that is, they convert the simple atmospheric form to slightly more complex inorganic molecules that can be used by plants. The best-known and most important of these nitrogen fixers are the bacterial mutualists that invade the roots of legumes, plants of the pea family. The plant develops root nodules (swellings) that apparently provide a suitable oxygen-free environment for the bacteria; the plant also supplies sugar. In return, it gets an abundance of fixed nitrogen, some of which is released into the soil where it can benefit other plants. Plants that do not have mutualistic bacteria absorb fixed nitrogen from soil and water. From plants, the nitrogen passes up food chains to animals. Various decomposing organisms break nitrogen compounds down again, and some, as I mentioned earlier, return nitrogen to the atmospheric pool.

If this complex nitrogen cycle were perturbed, the character of the atmosphere could change. For example, the atmospheric concentration of nitrous oxide could increase. That gas attacks the ozone layer (created from the oxygen produced by plants), which shields terrestrial organisms from dangerous wavelengths of ultraviolet radiation. A large increase in nitrous oxide therefore could result in an increase in the amount of ultraviolet radiation that reaches the earth's surface, a change which would be damaging in various ways to many forms of life. Apart from atmospheric effects, all plants and animals require large amounts of nitrogen for the construction of proteins; therefore, if the nitrogen cycle were to grind to a halt, we would too.

Natural ecosystems also help control and ameliorate the climate. They do so by influencing the flow of energy from the sun that is the engine that drives Earth's weather. That flow can be altered by changing the reflectivity of the atmosphere and the planet's surface, and thus the amount of solar energy that is absorbed. It also can be influenced by changing the degree to which the atmosphere can entrap solar energy that Earth's surface has absorbed—the "greenhouse effect." That effect is caused mainly by clouds, water vapor, and carbon dioxide (CO_2) in the atmosphere. All three tend to interfere with the reradiation of energy from the sun-warmed Earth's surface back into space. Since water vapor and CO_2 are largely transparent to incoming radiation from the sun, but absorb outgoing radiation in the form of heat, the net effect is to trap that heat from the sun near Earth's surface. Because of the greenhouse effect, the surface of Earth is some 60 degrees warmer on the average than it would be otherwise.

To illustrate the enormous influence ecosystems wield on these energy flows and thus on climates, consider transpiration by the lush vegetation of the Amazon basin. That huge expanse of dense forest recycles many times over the water that falls on it, producing the persistent cloud cover characteristic of the region. Clouds are great reflectors, intercepting incoming sunlight and bouncing it back into space. Thus the Amazonian ecosystem directly influences what becomes of the solar energy that reaches a significant portion of the surface of our planet. The clouds, reflecting sunlight and dropping rain, heavily modify the regional climate; they may even affect climates over the entire planet by influencing global atmospheric circulation patterns. Cutting of the Amazonian forest thus could possibly change global climatic patterns. And locally it might result in desertification, which already has occurred in deforested areas of northeastern Brazil.

Furthermore, changes in climate caused by alterations in ecosystems can be self-reinforcing. For instance, when large areas become desertified, the loss of vegetation itself changes both surface reflectivity and the rate of transpiration, and apparently leads to a further reduction of the already sparse precipitation. This seems to have happened in the Sahel.

Another way in which Earth's ecosystems control climate is by influencing the amount of CO_2 in the atmosphere. They do this through photosynthesis, respiration, and oceanic absorption, all of which are intimately involved in the carbon cycle and thus in determining the atmospheric content of carbon dioxide. As ecosystems are altered, that content will change. We know, for example, that the concentration of CO_2 in the atmosphere is now increasing significantly as a result of the burning of fossil fuels and probably also because of the clearing of tropical forests. Tropical deforestation adds CO_2 to the atmosphere both from burning and from decomposition of the wood. Part of the stock of carbon that was previously tied up in the biomass of the trees is returned to the atmosphere by both processes. Deforestation may also change the patterns of carbon flow by reducing the amount of photosynthesis, in which CO_2 is absorbed into the plants. Of course, the amount of photosynthesis taking place is not reduced if the forest is replaced by an equally productive community, such as fields of certain crops.

The precise role of forest clearing in the atmospheric buildup of CO_2 has been disputed, but there is absolutely no disagreement that the concentration has been steadily rising for many decades. And a significant increase in atmospheric CO_2 would be expected to have dramatic consequences: it would lead to a rise in average global temperatures by enhancing the greenhouse effect. A doubling of the concentration of CO_2 over its preindustrial level may occur by the middle of the next century, raising the average surface temperatures 2 to 4 degrees Fahrenheit, unless other factors counteract the CO_2 change. Knowledge of the response of ecosystems to increased CO_2 concentrations is essential for predicting the climatic changes that will be caused. Some responses may enhance and others may dampen the buildup. As an example of the latter, increased CO_2 could stimulate an overall increase in rates of photosynthesis and thus in uptake of CO_2.

The issue of global temperature change is of great significance because even a relatively small increase in the *average* global temperature could cause much larger temperature changes in some regions and substantially alter global circulation and precipitation patterns as well. Note that currently in the CO_2 debate much too much emphasis has been put on *average* global warming, and too little on the regional changes in temperature and rainfall that such an overall warming would be sure to cause. Indeed, some localities will probably become *colder* as the warmer atmosphere drives the climatic engine faster, causing streams of frigid air to move more rapidly away from the poles. Such changes could reduce agricultural yields for decades or more—a sure recipe for disaster in an increasingly overpopulated world. As University of California physicist John Holdren has said, it is possible that carbon-dioxide climate-induced famines could kill as many as a billion people before the year 2020.

Provision and regulation of fresh water are other crucial ecosystem services, supplied through control of precipitation, evaporation, and terrestrial water flows. Forest ecosystems are particularly important in supplying these services. In fact, the world over, forests and other vegetation types provide free flood control, soil-erosion control, and drought abatement. The loss of these services is illustrated annually by headline stories on American television when, during Southern California's rainy season, floods and mudslides plague hilly areas from which the chaparral has been burned off in the previous dry season. Similarly, disastrous spring floods in the Midwest and Southeast result not only from vast expanses of land bared for agriculture, but also from the drainage of swamps and other wetlands which formerly acted as reservoirs holding excess water.

A fine example of the role of forests in control of the hydrologic cycle is provided by the forest on the Virunga volcanoes in Rwanda, in central Africa. That forest, partially protected as the Parc National des Volcans, provides a home for the mountain gorillas. It makes up less than one-half of 1 percent of Rwanda's land area. Yet the forest acts as a huge sponge, absorbing rainfall and metering it out in streams that supply some 10 percent of that overpopulated nation's agricultural water. In the 1960s, the forest was two-thirds again as large as it is today, but in 1969 a large area of the park was cleared for the cultivation of the daisylike flowers that yield the natural pesticide pyrethrum. As a result, some streams dried up. And, it turns out, the pyrethrum growing was not an economic success (in contrast to the gorilla-based tourist trade, which is).

Expanding human populations are now placing enormous pressure on the Volcanoes Park. Rwandese peasants want to move in and clear land for agriculture. Should the rest of the park be deforested in order to accommodate a few months' growth of Rwanda's population, the result would be disaster for both the gorillas and the Rwandese—already the most crowded population in Africa and among the poorest in the world.

Aquatic ecosystems also purify water, decomposing wastes and cleansing water of many toxic substances and pathogens. This service is compromised or terminated whenever the quantities of wastes overwhelm the capacity of the system or when synthetic toxic substances are introduced. The decomposers have little or no evolutionary experience with the latter, and usually have no mechanisms for digesting them.

The generation and maintenance of soils, disposal of wastes, and cycling of nutrients are important, intimately interrelated functions of ecosystems. These involve a wide range of activities: the fragmenting of rocks by lichens and plants, the anchoring of soil by plants, the actions of decomposers and other organisms involved in cycling nutrients, including carbon, nitrogen, phosphorus, and sulphur. All these and a dozen or so others are critical in varying amounts to agriculture, grazing, and forestry. And, as noted above, the activities of some organisms are essential for maintaining the mix of gases in the atmosphere.

These vital services cannot be replaced on the required scale by human technology, and, in many cases, the knowledge of how to do it even on a limited basis is lacking. Indeed, even when people think they do it for themselves, in reality they have only harnessed other organisms to work for them; i.e., sewage treatment systems that employ decomposing bacteria to break down human wastes.

Natural ecosystems also control the vast majority of potential agricultural pests and carriers of human diseases. Remember, not very long ago mites were controlled by natural enemies everywhere—until humanity intervened with DDT. Remember, too, that *Opuntia* cactus is not a pest in South America, where appropriate herbivores are present to control it. In addition, natural ecosystems (both on land and in the sea) provide food for humanity and a vast array of medicines and substances useful for industry. They comprise an enormous genetic library of wild species and varieties from which we have already withdrawn the very basis of civilization—in the form of all crops and domestic animals, as well as innumerable useful products ranging from cotton, redwood, and rubber to wool, spices, vegetable oils, morphine, quinine, and digitalis. This library clearly has potential for yielding many more treasures, including better food crops for tropical areas, plants to grow in what amount to "gasoline farms" (petroleum is a geologically processed product of long-dead plants and microorganisms; some living plants produce similar chemical compounds today), and cures for cancer, heart diseases, and neurological disorders. Maintaining biotic diversity is one of the most important of ecosystem services. The accelerating destruction of this diversity should be of enormous concern to all of us.

You will recall that before I described the vital services supplied by ecosystems, I said that it is crucial to know how successional changes influence them. This is most easily seen by considering the impact of succession on the all-important nutrient cycles. Since the early 1960s, distinguished ecologists F. Herbert Bormann and Gene E. Likens, with their colleagues, have carried out landmark investigations of changes in a forested ecosystem in the valley of Hubbard Brook, a tributary of the Merrimack River in New Hampshire. One of their major goals was to understand the inputs, outputs, and internal cycling of nutrients in six small (ranging from 30 to 100 acres) contiguous watersheds—which, among other things, could yield information about the effects on nutrient cycles of forestry practices, as well as successional processes following clearing.

The Hubbard Brook ecosystem is perched on a foundation of waterproof rock, so the departure of nutrients in water could be easily traced; the only important pathways are streams. Precipitation could bring in nutrients (as for example, sulphur in acid rain), and wind or wandering animals could move them either in or out. Since winds in this area are rarely strong, and since the peregrinations of animals should pretty much even out (as much is carried into any given watershed as is carried out), the main nutrient flows could be traced by keeping track of those carried in by precipitation or out in running water. The movements of nutrients as gases (e.g., elemental nitrogen, sulphur dioxide) could not, however, be easily traced.

To determine the volume of inputs and outputs, devices for collecting rain and snow were stationed throughout the watersheds; and, where streams flowed out of them, small dams were built on bedrock to channel streamflow through measuring devices. Samples of both precipitation and stream water were chemically analyzed at regular intervals and the results combined with the data on input and output volumes, providing a general nutrient budget for the system.

The Hubbard Brook study has shown clearly that living and non-living parts of a terrestrial ecosystem combine to regulate the water, energy, and nutrients that flow through the system. But some of the most interesting observations were made on the effects of clear-cutting (chopping down the trees and letting them lie) in one watershed. In this case, there was minimum disturbance of the soil, but herbicides were applied to prevent regrowth of vegetation (and thus normal succession) for three growing seasons. Energy flows were immediately and dramatically modified. For example, the soil heated up substantially, partly because solar radiation previously reflected back into space or absorbed by the forest was able to reach the surface. But a second reason for the heating of the soil is less intuitively obvious than the lack of shading. When the Hubbard Brook forest was intact, an enormous amount of energy was used in transpiration—as huge quantities of water taken up from the soil by the roots of plants were released as vapor from the leaves and other parts. After the trees were removed, solar energy normally used for evaporation from leaf surfaces was free to reach the forest floor and raise its temperature.

The cessation of transpiration also increased the amount of water in the soil and the volume of summer streamflows. Much of the water that was previously drawn from the soil by the plants and evaporated from leaves moved instead over or through the soil to streams. And, naturally, the increased volume of departing water carried with it an increased volume of silt and nutrients from the exposed soil. The Hubbard Brook study and comparative research on a commercially clear-cut forest from which the logs were removed indicated that both systems suffered about an equal total loss of nutrients per unit area. Regrowth of plants in normal succession on the commercially cut area resulted in somewhat greater nutrient retention than at Hubbard Brook, where the herbiciding prevented uptake by new growth. But this superior retention was negated by the nutrients removed with the logs, nutrients that would be locked up in such places as the siding of houses or dining-room tables, or dispersed when waste wood or sawdust were burned.

In both clear-cut watersheds, nutrient losses were substantial, on the order of a quarter of the nitrogen and half of the calcium and potassium stored in the ecosystem before cutting. These studies showed that too-frequent clear-cutting or clear-cutting on steep slopes or thin soils could lead to progressive nutrient and soil loss that could permanently degrade the system and destroy its economic value.

These and many other results of the Hubbard Brook study have led to important recommendations for the management of forested ecosystems, such as trying to minimize disturbance of the forest floor in order to preserve the forest's capacity to undergo normal secondary succession. The mechanisms involved in nutrient flows such as those identified at Hubbard Brook are now major topics of research by ecosystem ecologists. Their studies are beginning to detail how such things as weathering of rocks and soils, erosion, nitrogen fixation, plant growth, and decomposition interact to produce and regulate these flows.

In an effort to understand the whole by detailed examination of the parts, ecologists have shifted attention toward the behavior of populations of the soil microorganisms that play such a critical role in nutrient movements. Peter Vitousek of Stanford has shown, for instance, that increased nitrate loss from ecosystems results from intensive forest site preparation, in which clear-cut sites are readied for replanting by removal of stumps, large roots, and the litter on the forest floor. This preparation reduces the availability of wood and other organic material that is necessary for the growth of populations of microorganisms that immobilize nitrogen in the soil and thus help to preserve its fertility.

When forests are being continuously managed for production of wood, the managers need to be alert to the ways that succession influences nutrient flows. Just as semiarid ecosystems today are continuously being converted to deserts, so are forests being continuously converted to unwooded ecosystems. This is an ominous trend, for, as I've already indicated, in addition to supplying wood, forest ecosystems perform many other services. And, sadly, in many parts of the world, when reforestation is attempted, the ecosystem services—such as running the hydrologic cycle and maintaining the genetic library—are not fully restored. A principal reason is that natural forests are replaced with monocultures of nonnative trees, which usually can at best provide a base for a relatively simple ecosystem far different from the original. In Uganda—which is slightly larger than Kansas, but now has proportionately less forest area than that state—stands of Australian eucalyptus have been planted where once native forests grew. One of the advantages of eucalyptus is that they are fast-growing. Unfortunately, however, they continuously shed bark and leaves, and relatively few other plants are able to grow in soil containing eucalyptus oil from the shed bark and leaves. Consequently, the ground cover within the stands is sparse, their water-metering function is presumably compromised, and the number of niches available to native plants and animals is greatly reduced.

It is obvious that clearing can disrupt the services supplied by a forest. But, at least in the north temperate zone, the chief threat to forest ecosystems is not physical but chemical. That threat can serve as an introduction to our last major section: how nutrient pathways have been mapped and the broad-scale changes humanity has made, or threatens to make, in ecosystems by altering the flows of nutrients and enlarging the flows of toxic substances.

In the north temperate zone, large amounts of sulphur and nitrogen oxides are injected into the atmosphere from factories, power plants, smelters, automobiles, and other sources. In the atmosphere, the oxides may undergo chemical reactions and be transformed into sulphuric and nitric acids, which then fall as acid precipitation; or sulphur dioxide may be deposited dry and be converted to acid on the leaves of plants or in the soil. Both the acids and sulphur dioxide can be toxic to trees and other vegetation.

The effects of these acid depositions are widespread. Acidification of ecosystems from such processes has already destroyed the fish and other animal populations in many hundreds of lakes in Scandinavia and northeastern North America. Tens of thousands of other lakes are threatened; about a third of the rivers in Nova Scotia that used to support salmon runs no longer do; and populations of salamanders are being exterminated over wide areas because the ponds they live in have become too acid.

In the forests of the Green Mountains of Vermont, acidification is probably responsible for reducing moss populations and thus interfering with their water-metering and erosion-control functions. The acids are also suspected of causing a decline of red spruce trees in the same area, especially in exposed locations at high altitude. Similar symptoms have appeared in forests of the southern Appalachians as well. In addition, John Harte of the University of California has found acid rain threatening montane ecosystems in the vicinity of the Rocky Mountain Biological Lab in Colorado, and during a visit to China he took a sample of rainwater in one small city that had the acidity of vinegar.

Damage caused by pollutants seems to be most severe in the forests of Central Europe. Large areas of forest there are dying, and scientists suspect that a major cause is either acidification and resultant changes in the chemistry of the soil, which interfere with water and nutrient uptake by trees, or the direct toxic effects of sulphur dioxide and another pollutant, ozone. Ozone is beneficial when it serves high in the stratosphere as a shield against ultraviolet rays, but when it is produced close to Earth's surface by the action of sunlight on nitrogen oxides and other pollutants, it is poisonous to plants.

According to soil scientist Bernhard Ulrich, acids can combine with nutrients such as magnesium and calcium and promote their leaching from the soil. The yellowing of the needles of many conifers is believed to be a sign of the magnesium deficiency that follows from such leaching. Scientists have also theorized that the acids are mobilizing (making available for uptake by plants) soil aluminum, which, though abundant, is normally held in harmless forms in the humus. The mobilized aluminum is toxic and will attack the roots of trees, interfering with the movement of water and nutrients into them and making them more susceptible to pathogenic organisms. Aluminum may also change the bacterial flora of the soil in ways that further increase the acidity.

Some scientists believe such changes in German forests and elsewhere may be irreversible on any practical time scale. Others are less pessimistic. At the moment, no one can really be sure precisely what is causing the damage to the forests, or exactly how it might be ameliorated. Ecologists agree, however, that the situation is potentially very serious, that it is produced by *Homo sapiens,* and that almost certainly it is related to the injection of those sulphur and nitrogen oxides into the skies.

The impact of acidification on aquatic systems, in contrast, is doubted only by scientists on the payrolls of the polluters. John Harte and his colleagues, for example, have used twelve-gallon microcosms to study the effects of lake acidification. Harte sets up each microcosm with lake water and some lake-bottom sediment carefully measured so that the water volume/sediment surface ratio is the same as in the original lake. Then he adds the appropriate acid at the rate it is being added in nature and watches the effects on dozens of species of plankton. The results, that various toxic metals are released from the sediment by the action of the acid, parallel very closely what happens in real lakes. If the ecosystem in the microcosm is degraded by the metals, the one in the natural lake will be also. The Harte group is elucidating the mechanisms behind the aquatic acidification disasters I described earlier, and developing systems that will permit relatively easy prediction of the fates of individual lakes now being subjected to acid deposition. The group is now also developing a soil microcosm, based on similar principles, in which to study acidification in terrestrial systems.

JOHN HARTE, UNIVERSITY OF CALIFORNIA

A researcher taking a water sample from an artificially acidified microcosm at the Lawrence Berkeley Laboratory. The microcosm has been stocked with water and sediment from Tenaya Lake in Yosemite, and the sample will be analyzed for acidity, trace metals (which are mobilized by acid), and the density of phytoplankton and zooplankton. The microcosm is in a plastic container surrounded by a steel jacket in which water regulated to the temperature of Tenaya Lake is circulated. The experiment, under the direction of John Harte, showed that Tenaya Lake sediments had little capacity to neutralize the acidity and contained a large amount of aluminum that was freed by the acidity. Acidification led to a major die-off of phytoplankton.

Useful as microcosms can be, their utility depends upon being able to check the results obtained in them with those obtained in natural systems. Unfortunately, tracing nutrient or toxin flows in natural aquatic systems is considerably more difficult than in terrestrial systems—since most elements can move around much more freely in water than in air or soil. In contrast, nutrients leached from soils as a result of deforestation, acid rains, or other disturbances simply flow from land into streams, rivers, and lakes.

It took, curiously enough, a side effect of the developments that made nuclear war possible to produce one of the most powerful tools now used for understanding nutrient pathways in natural aquatic systems. Shortly after the Second World War, nuclear reactors at the Oak Ridge National Laboratories began to make a radioactive isotope of phosphorus (^{32}P) available to ecologists. Isotopes are variants of an element in which the number of protons in the nuclei of the atoms is the same but the number of neutrons is different. Fortunately for ecologists, different isotopes of the same element are usually treated identically by natural systems—an amoeba is just as happy to incorporate ^{32}P into the compound adenosine triphosphate (ATP), which is crucial to its ability to use energy, as the much more common nonradioactive isotope of phosphorus, ^{31}P.

Because radioactive isotopes (radioisotopes) can be detected and their quantity measured easily, they are extremely useful as tracers. A radioisotope can be added to a natural system (such as a person's bloodstream or the water of a lake) and its movement within the system detected by the radiation it emits. The course taken by the radioisotope thus is a map to the normal route taken by that element in the system.

With the availability of ^{32}P, the decade of the 1950s became the dawn of the era of ecosystem physiology. By using ^{32}P, ecologists were able to analyze the phosphorus cycles of lakes. (G. Evelyn Hutchinson was a pioneer in this area, as in so many others.) It was discovered, for example, that older views of the cycling of phosphorus in aquatic systems were incomplete. It was long thought that there was a stately seasonal cycle in which supplies of phosphorus were gradually freed from sediments to accumulate in surface waters during the winter and then quickly consumed in spring by plankton "blooms," which died off and returned the phosphorus to the sediments. Summer plankton populations were believed to be limited by a lack of available phosphorus (and nitrogen).

Tracer work showed that part of the cycle was not stately at all—in fact, residence times of phosphorus in some small organisms (the time an average ^{32}P atom spent in the organism) were as short as a minute or so. It also showed that the sediments were not the most important source of phosphorus for the plankton. The most significant source was, rather, phosphorus that moved from the terrestrial surroundings into the lake water. There it undergoes rapid cycling through organisms in the water, flowing steadily through an organism-water-organism-water cycle downward into bottom sediments. In most natural lakes, relatively little of this phosphorus returns from the sediments and becomes available again to the plankton. The plants, animals, and microorganisms of the lake are thus dependent on input of phosphorus from the surrounding terrain.

If the normal nutrient flow is perturbed—say, if a load of phosphate-rich detergent is dumped into the lake—there will be a phytoplankton bloom as the previously phosphorus-limited populations quickly reproduce. But the whole system usually will return to its previous equilibrium as the phosphorus pulse moves through and is lost to the sediments. If a continuous input of detergent occurs, however, the now enriched lake will reach an equilibrium quite different from the previous natural condition. The change, called eutrophication, is considered undesirable—commonly, oxygen is depleted, killing valuable fish populations.

Fortunately, if the inputs stop, the lake will often return to approximately its previous state as the ecosystem purges itself of the unusually high phosphorus levels. This self-repair was seen in Lake Washington in Seattle. The lake was eutrophic in the 1960s, but when dumping of sewage (containing detergents and other nutrients) into the lake was stopped, the lake quickly recovered.

I've focused on phosphorus because ^{32}P has permitted relatively easy tracking of phosphorus movement through ecosystems. The more complicated cycling of the crucial element nitrogen has been more difficult to work out in detail, as it has no radioactive isotope. Something like ten times as much nitrogen as phosphorus is required by living systems, but nitrogen is an extremely abundant element. Phosphorus, in contrast, is quite rare and is much more often the nutrient whose shortage limits the growth of populations. One reason besides its relative rarity is that phosphorus does not exist in common compounds that are gaseous, while elemental nitrogen itself is a gas—the most abundant one in the atmosphere. All mineral nutrients tend to move from the land toward the sea, there to be eventually deposited in sediments. Once in the oceans, nitrogen can return to land through the atmosphere; phosphorus must wait for extremely slow geological processes to make the same journey.

Toxins as well as nutrients move from trophic level to trophic level in ecosystems, however. Poisonous substances introduced into the environment by humanity also are transported along both physical and biotic pathways. Many of the toxins are synthetics that are extremely persistent in the environment. As noted earlier, the microbial decomposers that ordinarily make their living breaking complex molecules into simpler ones have not evolved the capability of digesting many of the novel compounds created in organic chemistry labs. These compounds therefore can be a major source of disturbance of ecosystems, as when they poison animals that play important roles in delivering nature's pest-control service.

Of course, chemistry labs—even giant manufacturing plants—cannot turn out most compounds in large enough quantities to poison the vast volumes of water and soil into which they become mixed. There are, after all, some 60 octillion (60 followed by 27 zeros) tons of water in the oceans. If the production over the last decade of all synthetic pesticides—insecticides, herbicides, and fungicides—were stirred thoroughly into the oceans all at once, the concentration would be below 1/1,000,000,000,000,000—less than one part per quadrillion.

This point was made by a Nobel laureate in organic chemistry, Sir Robert Robinson. In a 1971 letter to the London *Times,* he claimed that compounds containing lead were not a threat to oceanic plankton because they would be so thoroughly diluted. He announced: "Neither our 'Prophets of Doom,' nor the legislators who are so easily frightened by them, are particularly fond of arithmetic . . ." and then proceeded to do some "simple arithmetic" (as he described it) of the sort I just did.

Unfortunately, that kind of arithmetic is just too simple. In his pompous pronouncement, Sir Robert missed the entire point. Rapid and complete mixing is a rarity in real physical systems, and for many substances that is precisely what does not happen when biological systems are involved. In particular, a phenomenon called biological amplification can produce concentrations of toxins in organisms many times higher than those found in the surrounding water, soil, or air. For example, lead is not evenly dispersed in the environment. It occurs in concentrations of about .02 parts per billion (ppb) in deep ocean waters, about .07 ppb in surface waters, and about 4 ppb (fifty to one hundred times the oceanic concentration) in average U.S. drinking water. But plants growing in cities may contain as much as 300,000 ppb in their tissues—15 million times the concentration in the ocean depths. One reason for this large accumulation in plants is that they readily take up, from both air and soils, tetraethyl lead, the form that is used as an antiknock additive in gasoline.

Plants, of course, have evolved the ability to extract certain substances differentially from soils; that is how they acquire the nutrients they need, and it is this capacity that causes them to concentrate some toxins. Similarly, sedentary animals, such as many marine invertebrates, also must extract what they need through selective absorption from the medium in which they grow, often with potentially unfortunate side effects for people who eat them. Oysters filter their food out of large volumes of seawater. Since they also live in shallow waters where pollution is the heaviest, they sometimes show up with astonishing concentrations of toxic compounds. Some oysters have been found to be loaded with 70,000 times the concentration of persistent pesticides found in the water around them.

Biological amplification thus can be due to differential movement of chemicals into organisms—either because the organisms actively take them up, or because the chemicals have a natural affinity for compounds found in the organisms' structure. Chlorinated hydrocarbons, of which DDT is the most famous example, very easily dissolve in fat but not in water. Since all organisms contain fat, there is a tendency for chlorinated hydrocarbons to leave water and enter plants, animals, and microbes, where they concentrate.

But there are other sources of biological amplification, and their effects can be seen in the great buildup of toxins in organisms at the upper end of food chains. In a famous study of a Long Island estuary, ecologist George Woodwell and his colleagues found concentrations of DDT under 0.1 ppm in water plants, 0.4 ppm in clams, and as much as 75 ppm in gulls. One reason for the higher concentration of DDT in organisms high on the food chains is that many of those organisms live for a long time and thus can accumulate more. Another is related to the energy loss at each step along the food chain as a consequence of the second law of thermodynamics. A typical animal may require ten pounds of food to replace or gain a pound of biomass, the remainder being excreted and used to run its life processes. Almost all of the DDT in the consumed food, however, will be retained in the animal's fat. The result is increased concentration.

The significance of this is that predator populations, high on food chains, are generally more vulnerable than herbivores to being poisoned by the accumulation of the toxic compounds that humanity injects into ecosystems. This is one additional reason that they are more subject to extinction than herbivore populations. Remember that predator populations are usually smaller than herbivore populations (and thus more prone to accidental extinction and the deleterious genetic consequences of small size), and, compared to herbivores, they have had little evolutionary "experience" with toxins and are less likely to have ready-to-go detoxifying mechanisms. Small wonder then that the standard result of heavy broadcast application of pesticides is the development of resistance in the pest and the decimation of its predators. And, as in the Cañete Valley cotton-pesticide disaster in Peru, it also leads to the promotion of new species of herbivores to pest status as their predators decline, reflecting general damage to the natural pest-control capability of the ecosystem.

How serious is the damage to ecosystem functions from physical and chemical assaults? About as serious as anything can be. Unless steps are taken to alter human behavior so that our species begins to tread lightly on the planet, pesticides, acid rains, forest clearance, desertification, and the other assaults against the ecosystems that support all life are bound eventually to lead to a collapse of civilization. But, while trends in environmental deterioration forecast disaster within, say, 50 to 150 years, there is one event that could produce a disaster with similar results in 50 to 150 days—a large-scale nuclear war.

A nuclear winter obviously would terminate most ecosystem services. As I explained earlier, darkness and depressed temperatures would halt photosynthesis, cutting off food chains at their base. Some plants would survive as seeds or underground parts, but productivity would essentially drop to zero over large areas of the planet and for a long time. Being utterly dependent on the plants, most animal life— including most human life—would perish.

Of course, a large-scale nuclear war would cause other severe impacts on ecosystems beyond cold and darkness. It is likely that the entire Northern Hemisphere would be bathed in a toxic smog created by towering infernos in plastic-filled cities, long-burning oil wells and coal stores, wildfires, and the like. Radioactive fallout would be much more intense than has been previously assumed—deadly enough to kill most of the pine trees and warm-blooded animals (which are more sensitive to radiation than most other plants and animals) in many areas. And once the atmosphere cleared, the returning sunlight would be enriched with dangerous wavelengths of ultraviolet light, because nitrogen oxides from thermonuclear explosions would have thinned the ozone layer in the upper atmosphere that normally screens out those wavelengths. This would damage the photosynthetic apparatus of plants and disorient insect pollinators—should any plants or insects have survived the cold and darkness.

Blast and fire themselves would also constitute massive insults to ecosystems, as would poisonous substances draining from places such as fractured storage tanks, broken transformers, the sumps of smashed automobile engines, and breached slime pits (the pollutant-laden tailing ponds of mines). Collapsed dams would release floods that would destroy outright the natural communities downstream, and silt-laden, possibly radioactive runoff from denuded land would exterminate much of the life in lakes, streams, and shallow marine waters. In short, a thermonuclear war could devastate ecosystems over much, or even all, of the planet, creating an extinction event comparable to or worse than those prehistoric ones discussed in Chapter 5.

Humanity has become a global force. One of many millions of species, *Homo sapiens* is now co-opting about a quarter of all the products of photosynthesis for its own use. Unfortunately, our species is also threatening destruction of much of the planet's store of organic diversity in a time span that, whether measured in minutes or in years, is the merest twinkling of an eye in relation to the eons that life has existed on Earth. It remains to be seen whether people will gain the ecological wisdom required to constrain their growing power before their own life support systems are totally destroyed.

EPILOGUE

THE FUTURE of the discipline of ecology appears bright—even if that of the biosphere does not. The time is ripe for breakthroughs in many areas in ecology. The next two decades might see not only the development of a general theory of what controls the size of populations but also an integration of knowledge of population processes into a theory of how ecosystems work. Ecologists have also come a long way in developing generalizations about the behavior of different aspects of the complex systems they investigate—density-dependent and -independent population regulation, foraging strategies, the evolution of sex and social systems, coevolution, resource partitioning, determinants of species diversity, food web structure, nutrient flows, and so on—so the ground has been prepared for further progress in these areas as well.

The darkest cloud on the horizon of ecology at present is the lack of appreciation of its importance by the general public. Many volumes have been published on the future of humanity, its economic system, and even the "population problem" with little or no consideration of the

288

overriding ecological issues involved in assuring a successful future for our species. Ignorance of the crucial importance of ecology is evidenced by the continuing low level of financial support for basic research in ecology relative to other areas of biological endeavor—particularly "biomedical research." The search to understand the molecular basis of carcinogenesis, with its potential for revealing ways to cure cancer, is something no one familiar with the pain and distress that cancer causes could fault. The same can be said for the efforts to understand and treat heart disease. Yet providing an environment that is less likely to induce these ailments, in many ways an ecological topic, probably holds the greatest hope for preventing both diseases.

More to the point, cancer and heart disease are comparatively minor problems compared with the struggle to understand the behavior of individuals, populations, communities, and ecosystems—all with the goal of helping to solve humanity's overwhelming problem of living peacefully on Earth within the constraints set by nature. For the vast majority of human beings, even cancer and heart disease are trivial concerns relative to those more closely related to fundamental research in ecology—the needs for safe, adequate, and sustainable provision of nourishment, energy, and materials for all people, and dependable control of such diseases as malaria, schistosomiasis, and dysentery.

Yet only one agency of the United States government, the National Science Foundation (NSF), supports basic research in ecology. The annual budgets of the NSF's sections of "Ecology and Systematics" and "Population Biology" amount to a total of some $34 million. In contrast, the budget of the National Institutes of Health (NIH) is now about $4,000 million—more than a hundred times as much. In their quite reasonable fear of the diseases that afflict them, middle-aged congressmen often fund biomedical research even beyond the level at which the money can be competently expended, sometimes wasting large amounts on the search for cures of diseases whose fundamental biology is not understood. The so-called "war on cancer" was a notorious example— millions more lives would have been saved if the same funds had been put into a "war on smoking." Indeed, as Jared Diamond points out, the existence of society would not be threatened if *no* progress toward a cancer cure were made over the next fifty years. But there *is* a deadline for solving humanity's ecological problems; for instance, less than a quarter of a century remains in which to take the steps necessary just to preserve the diversity stored in tropical rain forests.

If members of Congress understood the much more serious threat to themselves and their descendants posed by the ongoing destruction of their life-support systems, funding for those NSF programs in ecology would be immediately increased. To do this rationally, the funds would first have to be roughly quadrupled, then slowly moved toward dollar amounts of the same order of magnitude as those now given to the NIH. The money required, which would greatly enhance the chances of America surviving to the middle of the next century and beyond, could easily be found by eliminating some military programs—such as "Star Wars" —that reduce American security and, indeed, could help bring about the end of our society before the turn of the century.

Quadrupling the money now available to basic ecological research through the NSF would, I suspect, provide funds for all the investigations that could be done competently and efficiently by today's well-trained ecologists, evolutionists, population biologists, and taxonomists (most of whose research is now restricted by lack of financial support). A few million additional dollars per year would be required to initiate training programs that could greatly expand their ranks. It would probably take a crash program of almost a decade just to double the population of competent researchers. And that could be done only if society changed its priorities so that enough jobs in universities, government, and industry were available to absorb the additional scientists. At the moment, far more students wish to become ecologists than can find jobs as ecologists. Indeed, many of the bright young people who have already entered the field are having difficulty finding employment. What is clearly needed is a societal commitment to pay much more attention to basic research in ecology and the application of its findings to the solution of human problems.

Finally, perhaps $50 million a year, along with the increased support of individual investigators, could properly support the biological field stations around the world that are required as bases for doing research and training in ecology. Some of the best, like the Rocky Mountain Biological Laboratory and Fairleigh Dickinson University's superb West Indies Laboratory, operate on shoestrings. Others, like Tanzania's Serengeti Research Institute, are now essentially nonfunctional for lack of funds. And the critical training programs of the Organization of Tropical Studies in Costa Rica are chronically short of support.

In addition to supporting basic research, an immensely larger commitment of funds and talent to the preservation of biological diversity —especially through the protection of tropical forests—is essential immediately. How to use these funds and talent is beyond the scope of this discussion, with one small exception. The effort to comprehend and catalog that diversity is centered in poorly funded natural-history museums, staffed by all-too-often badly paid taxonomists. The United States should support its museum operations at a much higher level and at the same time cooperate in a massive international effort to stem the tide of extinctions that threaten Earth's biota and the future of our own species.

Even though there is no sign yet that such commitments are forthcoming, at least ecologists themselves are finally beginning to push for them. The recent studies of the long-term, worldwide effects of nuclear war, in which a large number of ecologists and evolutionists collaborated, was, I think, indicative of a growing realization that the time has come to make our voices heard in the councils of power. The Ecological Society of America recently opened an office in Washington, D.C. It is being run by Elliot Norse, a young scientist who understands both the discipline and the political challenges facing it. Many leaders of ecology in my generation—Jared Diamond, Tom Eisner, Bob May, Hal Mooney, Peter Raven, Ken Watt, Ed Wilson, George Woodwell, and others—have been "going public" with their concerns. They have the encouragement of distinguished senior scientists like Charles Birch, G. Evelyn Hutchinson, Ernst Mayr, Charles Michener, and Eugene Odum, and the support of the large numbers of younger scientists who now are making ecology an ever more exciting field.

There even seems to be some movement in the right direction in government agencies and private foundations. Both groups, for instance, have a growing interest in funding research in the applied subdiscipline of ecology, conservation biology, whose explicit goal is to develop the scientific tools necessary for preserving Earth's rapidly dwindling treasure-house of biological diversity.

Time, however, is growing short. Nature's machinery is being demolished at an accelerating rate, before humanity has even determined exactly how it works. Much of the damage is irreversible. To protect and (to the extent possible) repair the machinery will require many more scientists delving into its complexities, as well as a much deeper appreciation by the general public and decision makers of both its importance and its basic design. More young people must be given both the stimulus and the opportunity of making careers in ecology. And all of humanity must gain more empathy with nature. Since the future of civilization depends on nature, only avoidance of a nuclear holocaust should have equal priority with the movement toward ecological understanding.

APPENDIX A

Electrophoresis

Basically, electrophoresis involves grinding up an organism, or part of an organism, in an appropriate solution. The resultant mess is sopped in blotting paper and stuck into the equivalent of a pan of gelatin, or "gel." The gel is then placed in an electric field that causes the proteins in the sample, which carry electrical charges, to migrate through the gel. Proteins are made up of chemical building blocks called amino acids, and the twenty different amino acids show differences in their electrical charges. Thus the rate of migration of a protein molecule (the smallest unit of the protein) through the gel is determined both by its size and by its electrical charge, which are functions of its amino-acid composition.

But how can one discover which protein has moved where? Most of the proteins studied by electrophoresis are enzymes—which are organic catalysts that affect the rates of the life processes of every organism. For example, the enzyme hexokinase binds to glucose, a form of sugar, and facilitates the addition of a chemical complex containing phosphorus to the sugar. The result is a compound called glucose 6-phosphate. In this example, glucose is called the substrate and glucose 6-phosphate is the product. To discover where molecules of a given enzyme have migrated in a gel, it is only necessary to know the substrate upon which they act and to have a technique of staining that will reveal the presence of the product. When a gel is treated with the proper substrate and stain, the positions of the enzyme molecules are revealed as stained bands. And if different forms of the enzyme are present, some of the bands normally will have traveled farther than others, indicating a difference in the amino-acid building blocks that went into them.

How does all this connect to genetic variability? It turns out that the more variation there is in the distances migrated (in a given time), the more genetic variability is present. Many thousands of different proteins can be constructed, using various configurations of the twenty amino acids. The kinds and sequences of amino acids that go into proteins are determined directly by the DNA; that is, by the genes of an organism. That, in fact, is what DNA is all about—programming the assembly of proteins. So the variation in migration distances in stained bands in electrophoresis gels represents variation in the genetic makeup of the organism.

If you took biology in high school or college, you may have tried some simple experiments in Mendelian genetics in which fruit flies with, say, normal red eyes were crossed with others with white eyes. Then the frequencies of normal- and white-eyed individuals were recorded in several generations of their offspring to work out the system of Mendelian inheritance. Exactly the same kinds of experiments can be done with fruit flies that have variations of a given enzyme that migrate in a gel at different rates—and in most cases it can be shown that those differences are under the same kind of simple genetic control as are the eye colors.

APPENDIX B

Resolution of the Neutrality Controversy

In addition to mutation and selection, two other important forces are operating on the collective genetic endowment (gene pool) of a population. The first is migration—individuals carrying genes into or out of the population. Clearly, this "gene flow" can have a significant impact on the genetics of a population. In fact, if genes are neutral, a very small amount of intermigration can keep their frequencies the same in two populations. This makes knowledge of gene flow extremely important in attempts to determine from geographic patterns whether observed genetic variation in populations is neutral. If populations have similar gene frequencies and are totally isolated, it could indicate that selection is maintaining those frequencies. If, however, there is even a small amount of gene flow between them, then the variation could be neutral and their similar frequencies maintained by the migration.

The second important factor is random changes in the genetic composition of a population resulting, among other things, from accidents that help determine who does and who does not reproduce.* Random changes will occur with or without selection.

* A major cause of random changes is sampling error that is inherent in the way the genetic mechanism functions—for instance, the sperm and eggs that unite

The importance of random changes is closely related to the size of the population, as can be seen from a simple example. Suppose there were a population of four bugs, one of which possessed a novel gene. If a cow happened to step on one of the bugs, the chance of losing the gene from the population would be 25 percent. On the other hand, suppose there were a population of 1,000 bugs, of which 250 carried the novel gene, and cows accidentally stepped on and killed 250 of the 1,000 bugs. What is the chance that, by accident, they would all be the bugs carrying the novel gene? The chance is extremely small— for practical purposes, the possibility can be ignored.

Various random events cause the proportions of different kinds of genes to rise and fall aimlessly—to drift up and down. For this reason, it is often said that random events cause genetic drift. Thus in small populations chance can play a very significant role in shaping the gene pool; in large populations it usually cannot.

Unlike genetic drift, the other evolutionary factors change the proportions of genes in a systematic way. Selection favors one kind of genetic information over another. Mutation changes one kind of genetic information into another. Migration adds or subtracts one kind of genetic information to or from the population.

The major problem confronting the people who are trying to resolve the neutrality controversy is sorting out the effects of selection from those of mutation, migration, and drift. Mathematical models created to do this—to investigate the behavior of individual genes and groups of genes under selectionist and neutralist assumptions—have met with varying degrees of success.

My Stanford colleague Marc Feldman, for instance, is developing a theory focused on one crucial aspect of the problem—how one can judge whether an enzyme gene that appears to be under selection really is. It might instead be a "neutral" gene that is just "hitchhiking" on another gene that *is* being selected and which happens to be physically close to the enzyme gene on the chromosome (such physically close genes tend to be transmitted together). Marc's work lies in the domain of population genetics, a field that has already developed a large body of theory quantifying what will happen to the frequencies of genes in populations that are subjected to varying regimes of mutation, migration, selection, and drift. But developing models useful in settling the neutrality controversy by the analysis of observed patterns of genetic variation has proved difficult.

There are also two approaches to the problem based not on theory but on experiment and observation. One, exemplified by the work of another Stanford colleague, Ward Watt, has been to look at the ability of slightly different variants

---

to form a new generation are only a sample of those that might have been produced and a sample of those that were produced that might have united. Just as ten flips of an honest coin can, because of sampling error, produce six heads and four tails (a deviation from the expected five each), so sampling error can cause the gene frequencies in the offspring generation to differ from those of the parental generation. This and what follows in the text of the appendix is simplified, but represents the gist of a complex subject.

of one enzyme to control the rates of chemical reactions in an organism's metabolism and at the conditions under which those variants occur in natural populations to see if some relationship exists between them. Ward has been building an impressive case that the variations in some butterfly enzymes can be accounted for by their performance at different temperatures. Butterfly populations that live in colder environments tend to have enzyme variants that, in laboratory preparations, are more efficient at cooler temperatures than are the variants that predominate in the butterfly populations in warmer locales. This, of course, indicates that selection pressures in cooler and warmer environments are different, and that differential reproduction of the genotypes did indeed produce the different varieties of enzymes; the variation being studied is thus not neutral.

Ward has made remarkable progress in elucidating how evolution shapes the details of an organism's metabolism. His investigations have required an extremely time-consuming and meticulous combination of field and laboratory experimentation. But even such careful research as this has not been able to resolve the neutrality controversy. The demonstration that variants of one enzyme in a single organism are under selective control is but one of a great many bits of data required to settle the issue. Even the most orthodox neutralists agree that in some cases selection *must* operate on enzyme systems. Only when geneticists know what happens to dozens of enzymes in hundreds or thousands of creatures can there be a general answer to the neutrality question. Unfortunately, very few biologists have Ward Watt's combination of talents and interests. So Ward's approach to the study of enzyme variation, while it is producing information of great general interest to evolutionists, holds little hope of a quick solution to the controversy.

Another approach to resolving the dispute has been to analyze many patterns of variation in enzymes across geographic areas or through time without studying the properties of the individual enzymes. That is the approach to the problem that our research group and many others have taken. Our group has an enormous advantage in that we know a great deal about the ecology of the checkerspot butterflies we are studying and especially about the ability of individuals to move between populations and transfer genes in the process. We know that, in many situations, there is no significant migration between checkerspot populations. We are therefore able to eliminate consideration of that one key variable that could affect gene frequencies and concentrate on trying to differentiate between the effects of selection, mutation, and drift.

In spite of this advantage, the results have not been encouraging. So far, after a great deal of effort, we have been able to determine, by a process of elimination, only that selection must be involved, either directly or indirectly (via hitchhiking), in determining the patterns we see in a small number of enzymes in this one kind of organism. Our problem is analogous to Ward's—while we are learning interesting things, the logistic effort required to add even a bit of data to the solution of the neutrality controversy has been enormous.

APPENDIX C

Statistics, Null Hypotheses, and Checkerboards

The chance of an honest coin coming up heads ten times in a row by chance is calculated as follows. At each toss, the chance of a head is $\frac{1}{2}$ (or 0.5). Since the coin has no memory, statisticians would say the tosses are "independent," and the chance of two heads in a row is $\frac{1}{2}$ times $\frac{1}{2}$, or $\frac{1}{4}$, and of ten heads in a row $\frac{1}{2}$ times itself nine times (or $.5^{10}$). The answer is $\frac{1}{1024}$, or about .001. That is, the probability of the event just observed, ten heads in a row, happening just by chance is slightly less than one in a thousand. Such a result is unusual, to say the least, and if it happened, it would probably lead one to examine the coin. On the other hand, if a sequence of ten flips is tried a couple of thousand times, one would expect about two sequences of all heads (and about two of all tails) to show up.

The level of probability chosen as significant (that is, how small the probability has to be before chance is rejected as the cause of the observed phenomenon) is largely a matter of judgment. For example, suppose a test were done on two groups of human volunteers to decide whether an expensive drug reduced the severity of colds. The group that received the drug, on the average, recovered from colds more rapidly than the control group that received only a placebo. But, of course, two groups of people would not get over colds in exactly

297

the same time even if they were given no drugs at all. A statistical test must be applied to determine the probability that the drug actually influenced the outcome—in other words, a test of the so-called "null hypothesis" that the difference was due to chance alone. And one would want to be very sure that the group that received the drug recovered not just by happenstance, but because the drug worked. Otherwise, people might waste a lot of money on an ineffectual cure. Therefore one might wish to demand a probability of less than .001 that chance alone was responsible before rejecting the null hypothesis that the drug was no different from the placebo. The cost of rejecting the null hypothesis (deciding the drug was effective) when the hypothesis was actually true (drug ineffective, difference caused by chance) would be high.

In contrast, if one were using mice to test defensive chemicals from plants to see whether the chemicals had any efficacy against mouse cancers, one might accept a one in twenty (.05) chance of mistakenly thinking one drug was better than the "no drug" treatment. It would clearly be a mistake not to investigate every compound that showed promise, and the cost of wrongly rejecting the null hypothesis would be relatively low—simply some further investigation before deciding the chemical really had no anticancer potential.

How would one construct a null hypothesis about the distribution of birds on islands for testing whether checkerboard distributions were caused by biological interactions? One way the Simberloff group has done this was to use a computer to generate all the possible arrangements of birds on islands, using the actual number of bird species in the fauna, the actual number of islands, and placing on each island the number of birds actually found there. In a simple example, suppose there were just three species, 1, 2, and 3, and two islands, A and B. Suppose the actual fauna was species 1 on island A, and 2 and 3 on B (A-1; B-2,3). The other possible arrangements under the rules presented above are (A-2; B-1,3) and (A-3; B-1,2). The arrangement of the actual fauna plus the other possible arrangements make up a "null distribution"—the array of arrangements possible under the rules. One can then see how likely the actual arrangement is compared to the possible ones.

In the above case, it would be difficult (at least on the basis of the observed geographic distribution alone) to conclude that either species 2 or 3 (or the two together) competitively exclude species 1 from island B. After all, if the birds were randomly allotted to the islands (that is, the distribution were due to chance), the observed combination would occur in a third of the cases. But, as you will see in the text, there are serious problems with the basic rationale of this test.

FURTHER READING

A more technical treatment of the material in this book, and references to the scientific papers on which it is based, will be found in the forthcoming text *The Science of Ecology* (New York: Macmillan) which I wrote with Jonathan Roughgarden. It has the same general structure as this book and, naturally, the same general prejudices. *Ecoscience: Population, Resources, Environment* (San Francisco: W. H. Freeman, 1977) written with Anne Ehrlich and John Holdren, and *Extinction: The Causes and Consequences of the Disappearance of Species* (New York: Random House, 1981) by Anne and myself, deal with the interface of technical ecology and environmental issues. Both books are heavily documented, but are designed for a lay audience.

To get some idea of the diversity of approaches to ecology you might wish to consult one or more of the technical books listed below. They are only a sample of the fine volumes available, but a rather representative one. Unless otherwise noted all require only an elementary knowledge of biology for one to get a great deal out of them, although some knowledge of algebra and the calculus would enhance understanding of parts of most. One other book I should mention: E. O. Wilson's *Biophilia* (Cambridge, Mass.: Harvard University Press, 1984) is a wonderful essay arguing that human beings have a natural empathy for other living organisms. It is the best description I have read of a working ecologist's feelings about the living world.

Andrewartha, H. G., and L. C. Birch. 1954. *The Distribution and Abundance of Animals*. University of Chicago Press. A classic treatise on population ecology. For many biologists of my generation, this comprehensive volume was a breath of fresh air in a discipline gone stale.

———— 1984. *The Ecological Web*. University of Chicago Press. Designed to update their classic, and containing a sure-to-be-controversial treatment of basic ecology, since it questions the sorts of theoretical treatments introduced by Hutchinson and MacArthur. Whatever one's opinion of that, this is a fine source book of population ecology, containing a wealth of detail.

Brown, J. H., and A. C. Gibson. 1983. *Biogeography*. C. V. Mosby, St. Louis. A broad, modern treatment of the subject.

Dobzhansky, T., F. J. Ayala, G. L. Stebbins, and J. W. Valentine. 1977. *Evolution*. W. H. Freeman, San Francisco. With Futuyma's text (see below) one of the two best recent treatments of the neo-Darwinian synthesis.

Elton, C. 1927. *Animal Ecology*. Science Paperbacks, Methuen and Co., London. This pioneering text by Elton, who is often considered the father of modern ecology, is available in paperback. In it, Elton wrote (p. vi) that "ecology is the branch of zoology which is perhaps more able to offer immediate practical help to mankind than any of the others, and in the present rather parlous state of civilisation it would seem particularly important to include it in the training of young zoologists."

Futuyma, D. J. 1979. *Evolutionary Biology*. Sinauer Associates, Sunderland, Mass. A fine textbook, presenting the modern consensus on the process of evolution.

————, and M. Slatkin, eds. 1983. *Coevolution*. Sinauer, Sunderland, Mass. A multi-authored volume with consistently excellent individual contributions.

Harper, J. L. 1977. *Population Biology of Plants*. Academic Press, London. *The* volume on this topic; the "Andrewartha and Birch" of the plant world.

Hutchinson, G. E. 1978. *An Introduction to Population Ecology*. Yale University Press, New Haven. Scholarly and brilliant—the footnotes are wonderful. Hutchinson's approach is very different from, and complementary to, that of Andrewartha and Birch.

Krebs, C. J. 1978. *Ecology: The Experimental Analysis of Distribution and Abundance*, 2nd ed. Harper and Row, New York. A data-rich text with a bias toward population ecology in the tradition of Andrewartha and Birch.

Krebs, J. R., and N. B. Davies, eds. 1978. *Behavioral Ecology: An Evolutionary Approach*. Sinauer, Sunderland, Mass. This multi-authored, rather controversial volume deals with an area that tends to be neglected in most ecology texts.

MacArthur, R. H. 1972. *Geographical Ecology*. Harper and Row, New York. This landmark text, tightly written, and built around a structure of mathematical theory, provides a good introduction to how MacArthur viewed the world—and bears the stamp of his mentor, G. Evelyn Hutchinson.

Mayr, E. 1963. *Animal Species and Evolution.* Harvard University Press, Cambridge, Mass. My favorite (and I think the most accessible) of the classics of the neo-Darwinian evolutionary synthesis.

Milkman, R., ed. 1982. *Perspectives on Evolution.* Sinauer Associates, Sunderland, Mass. An excellent companion volume for Futuyma's text. The chapters by Stephen Jay Gould and Guy Bush provide insight into the controversies over mechanisms of speciation.

Odum, E. P. 1971. *Fundamentals of Ecology,* 3rd ed. W. B. Saunders Co., Philadelphia. A comprehensive text taking an "ecosystem approach"—by a great pioneer of that approach.

Ricklefs, R. E. 1979. *Ecology,* 2nd ed. Chiron Press, New York. Readable, balanced, and encyclopedic. It is rumored that a new edition of this popular text will be published soon.

Roughgarden, J. 1979. *Theory of Population Genetics and Evolutionary Ecology: An Introduction.* Macmillan, New York. This is the best text for those mathematically inclined who wish to understand basic theory. Some calculus is required of the reader.

Soulé, M. E., and B. A. Wilcox, eds. 1980. *Conservation Biology: An Evolutionary-Ecological Perspective.* Sinauer Associates, Sunderland, Mass. Every serious environmentalist should own this paperback volume. Its publication marked the genesis of conservation biology as an integrated discipline.

Southwood, T. R. E. 1966. *Ecological Methods.* Methuen and Co., London. A look at this classic work will give you a good idea of how ecologists actually go about their work.

Vermeij, G. J. 1978. *Biogeography and Adaptation: Patterns of Marine Life.* Harvard University Press, Cambridge, Mass. A stimulating and idiosyncratic book on marine ecology and evolution.

Whittaker, R. H. 1975. *Communities and Ecosystems,* 2nd ed. Macmillan, New York. A relatively brief paperback ecology text that, like Odum's book, has a strong emphasis on ecosystems.

Wilson, E. O. 1975. *Sociobiology.* Harvard University Press, Cambridge, Mass. An absolute gold mine of information for anyone interested in the behavioral aspects of ecology.

ACKNOWLEDGMENTS

First and foremost I owe thanks to Anne Ehrlich and Jonathan Roughgarden. Anne slaved with me over the text, reading it more times than I can remember, and making hundreds of excellent suggestions and additions. Jon also read and criticized the text, but more importantly he has influenced my thinking about ecology enormously, both in years of conversations and in the course of writing our textbook, *The Science of Ecology,* which covers much of the material in this book at a much more technical level.

Other colleagues in Stanford's Department of Biological Sciences have also helped with critiques and suggestions. They include David Dobkin, Marcus W. Feldman, H. Craig Heller, Richard W. Holm, Harold A. Mooney, Dennis D. Murphy, Kenneth H. Naganuma, Peter M. Vitousek, Darryl Wheye, and Bruce Wilcox. Other people who read and criticized all or part of the manuscript were Kenneth B. Armitage (University of Kansas), Steven Grunow (Harvard), Marnie Hagmann (Simon and Schuster), John Harte and John Holdren (University of California, Berkeley), Cheryl Holdren (Rocky Mountain Biological Laboratory), and Sally and Robert Ornstein (Institute for the Study of Human Knowledge). I am deeply indebted to all these people, but especially to Sally Ornstein and Darryl Wheye, whose help in trying to make the text accessible to those without technical training was invaluable (even though I may have failed in spite of their efforts).

One of the many advantages of being on the faculty at Stanford is access to the best-run biology library in the world. As usual the staff of the Falconer Library, under the direction of Beth Weil, was enormously helpful to me. Beth, Claire Shoens, Zoe Chandik, Judy Levitt, and Joan Dietrich not only dug up numerous obscure references, obtained interlibrary loans, found mislaid books and journal articles for me, but did so cheerfully, never running for cover as I approached. Steven Masley Xeroxed many pages from journals flawlessly and promptly. All of these people went beyond just "doing their jobs," and I appreciate it.

In addition to reading the manuscript, Darryl Wheye was extremely helpful in organizing the literature and helping me to deal with various computer problems that popped up in the course of word processing the text. Robert C. Eckhardt and Bob Bender at Simon and Schuster and Ginger Barber of the Virginia Barber Agency, besides providing excellent and repeated critiques, were helpful to me in more ways than I can recount. Robert, a trained ecologist, went far beyond the call of duty in helping me to deal with highly technical subjects both simply and accurately. Patricia Miller did a fine job of copy-editing the manuscript.

I also owe thanks to numerous colleagues in ecology and evolutionary biology who have done the scientific research on which this book is based. Some of them are mentioned by name, but many more are not. I hope the latter will understand that a book for laypeople cannot be documented as a text or journal article should be, and that in most cases their work will be credited to them in *The Science of Ecology*.

Much of our group's work discussed in this book was funded by the National Science Foundation and the Koret Foundation of San Francisco. Their support is very much appreciated, as is that of the Rocky Mountain Biological Laboratory in Crested Butte, Colorado.

Finally, I remain deeply in the debt of LuEsther, whose friendship has meant more to me than I could ever tell her.

INDEX